T0136444

PHILOSOPHICAL

INSTRUMENTS

PHILOSOPHICAL INSTRUMENTS

MINDS AND TOOLS AT WORK

Daniel Rothbart

FOREWORD BY ROM HARRÉ

UNIVERSITY OF ILLINOIS PRESS

Urbana and Chicago

∞ This book is printed
on acid-free paper.

Library of Congress Cataloging-in-
Publication Data
Rothbart, Daniel.
Philosophical instruments : minds and
tools at work / Daniel Rothbart ;
foreword by Rom Harré.
p.   cm.
Includes bibliographical references and
index.
ISBN-13: 978-0-252-03136-6 (cloth : alk.
paper)
ISBN-10: 0-252-03136-9 (cloth : alk. paper)
1. Knowledge, Theory of.
2. Experimental design.
3. Science—Philosophy.
4. Scientific apparatus and instruments.
I. Title.
Q175.32.K45R68   2007
501—dc22   2006025174

Dedicated to Florence H. Rothbart
and Harold A. Rothbart

CONTENTS

FOREWORD

Rom Harré

For more than a century, the philosophy of science was almost exclusively concerned with the discourses of science, knowledge in verbal form. The role of the laboratory, for example, in the analysis of material substances, and the role of the expedition to some site of interest, such as the Galapagos Islands, were reduced to mere neutral sources of propositions. An experimentalist was just reporting the results of experiments and observations. Even in the new sociological approach to the sciences, philosophers took the instruments and apparatus to be devices for producing text. What of the equipment itself? Is it rightly taken for granted, as a neutral substance to all science, of no intrinsic interest?

In this study, Daniel Rothbart has presented the first mature and definitive account of the nature of the material things with which scientists engage the world and of the immanence in equipment and the materials of which it is made. The stuff of the laboratory expresses and realizes philosophical stances to the world. The last decade has seen a flurry of publications on various aspects of these topics, but until now, a comprehensive treatment has been lacking. The key that opens the way to an understanding of how material things can be used to create knowledge is the realization that experimentation and observation are skills. However, to create an instrument or an apparatus is itself a skilled procedure, involving not only the skills of the designer but also the skills of the instrument maker to transform designs into material objects. The story of the development of a successful air pump illustrates the interplay between innovation in design and expanding possibilities of experimental programs in which the instrument plays an indispensable role. Once in being, an instrument or apparatus is a tool for use. It can be used skillfully or unskillfully. Mere prescription of a routine for using a tool is not enough to ensure that it is effectively put to use.

In the uses of equipment, be it on the massive scale of a tokamak to be used for nuclear fusion or a pair of binoculars to be used by an ornithologist, instruments depend on something yet more fundamental, intuitions of the material possibilities of the very stuff with which engineers create them. Design is useless unless there is material stuff in which it can be realized. To master the making of instruments is also to have a mastery of the potentialities of existing materials. However, it is also necessary to have a vision of how new material might be developed, with as yet unperformed tasks in mind. Glass vessels will not do as containers for fluoric acid, but containers of fused silica will serve.

How much of the natural world will be available to human exploration? Microscopes, in skilled hands, enhance the powers of vision, but they also make manipulations of hitherto unimaginable sensitivity possible. Rothbart brings out with copious illustrations the importance of the insight that access to what cannot be observed is achieved not only by enhancing the senses, creating "microscopical eyes," but also, and above all, by enhancing the power of the hand to manipulate the fine structure in the material world to bring to light hitherto unrealized powers and capacities.

In these studies, Rothbart opens out the concept of a machine in several directions. The world as a natural machine is a familiar metaphor. However, specimens as machines—that is, a material system the powers and capacities of which can be skillfully and insightfully exploited in gaining an understanding of the natures of things—is a further and exciting enlargement of the metaphor. The root situation in the natural sciences is always something like this: Here is a machine. What can we make it do? Even those machines we designed ourselves can surprise us, revealing capacities that were not in the minds and hands of those who first put them together. Even more surprising are the machines we find in nature.

I have to confess that I love instruments and apparatus just as beautiful things. Appreciation of the form/function relation is not all there is to this aesthetic thrill. That which is well made has its own beauty just by virtue of that aspect of it alone. What is well made is likely to work well, though notoriously that is not universally true. Rothbart shares this fascination with the aesthetics of instruments and apparatus. In museum collections there are decorated instruments from the seventeenth and eighteenth centuries. Tooled leather and engraved silver stands enhance many a microscope. The aesthetics that is celebrated and analyzed in this study of the beauty of instruments and apparatus as such is intrinsic to it. It is not a mere superficial tricking out in "buttons and bows." It is more than the display of the economical realization of function.

This book is a major contribution to a burgeoning field of the greatest philosophical interest. In these studies, we see very vividly how philosophy has grown away from the logicism of the Russellian era. The studies of instruments and apparatus has led beyond the confines of discourse to the material activities of human beings as they use their hands to reach beyond the limits of their eyes in search of knowledge of the world in which we live.

PREFACE

While writing this book, I often returned to the following thought: some scientific practices are good (efficient, productive, promising, or valid), and others are bad. With this simple distinction, philosophical debate begins. In centuries of debate about empirical knowledge, almost every kind of practice related to science (good and bad) has been probed, scrutinized, and dissected. Consider the practices of experimental researchers. Rationalist philosophers tend to trust the scientists' pronouncements of successful research as sources of knowledge about real-world events. Most relativists of the sociological variety respond with skepticism about such pronouncements as they seek to expose social and cultural forces that can influence researchers and contaminate results. According to this view, experimenters often exploit the use of such honorific terms as "genuine knowledge" and "real-world process" in propaganda campaigns for personal glory or power.

Are experimental practices sources of "fact" or "fraud"? This controversy turns on a pivotal aspect of research practice. The topic of instrumental skill, pervasive in current research, demands its own principles, which are not reducible to traditional categories of empirical knowledge. For any instrumental skill, a human agent deploys tools and anticipates the effects on some segments of the environment. Skill demands both practice and thought. In this work, I examine how instrumental skills "carry" ideas about the idealized relationships between experimenters' actions and specimens' properties. We can read in design plans for instruments philosophical commitments about the technological interactions between experimenter and specimen. Designers of laboratory instruments define "good" practice from a distillation of the past and an idealized vision of promising results in the pursuit of knowledge. As tools blend with minds in the skilled use of laboratory instruments, philosophical ideas about knowledge-inquiry are deployed and tested. A robust philosophy of experimentation emerges by reflecting on the design plans for such mediating technologies.

In writing this book, I drew upon conversations with and the writings of

Jim Bennet, Tom Dietz, Phil Dowe, David Gooding, Tim Ingold, Joachim Schummer, and Debra Warner. Ian Hacking's work became a source of inspiration, especially his insight into the synthesis of knowledge and action in laboratory research. Ideas coming from the pen—and word processor—of Rom Harré find their place throughout my philosophical explorations. In Harré's work, realism is rescued from the onslaught of relativism through a blend of metaphysics and material practice. In the present work, I examine how certain technical skills, as defined through design plans for instruments, reveal metaphysical commitments about inquiry.

Insight and inspiration have also come from colleagues at my home institution, George Mason University, in particular James Fletcher, Wayne Froman, Emmett Holman, and Ted Kinnaman as well as members of the International Society for the Philosophy of Chemistry, most notably Eric Scerri, Michael Akeroyd, Joseph Earley, Klaus Ruthenberg, Davis Baird, and Eva Zielonacka-Lis. The meetings of this society provided an excellent forum for refining many of the themes presented in this book. Additionally, I am indebted to Irmgard Scherer for her expertise in Kant, John Schreifels for his instruction on spectrographic techniques, and Suzanne Slayden for her meticulous research on absorption spectroscopy. Outside of academic circles, I gained valuable information from Alane Keller, Thom Larson, and Eric Van Der Veer.

Also, from my home institution, I thank the chair of the Department of Philosophy, the dean of the College of Arts and Sciences, and the provost for generously supporting my participation at various conferences and meetings, where many of these ideas were first voiced. Resources associated with the 2000–2001 Fenwick Fellowship award contributed to my research activities. Apologies must be given to my graduate students who were subjected to the initial expression of some ideas in this book and who responded graciously and with appropriate requests for clarifications and illustrations. I thank Matt Scanlon and Heather Tinsley for their excellent editorial reviews, including the addition of creative phrases. Kerry Callahan, acquisitions editor of the University of Illinois Press, gave support and encouragement throughout this project. I especially want to thank the two anonymous reviewers from the Press for their meticulous examination of the first draft and for pushing my ideas forward, demanding more detail of my arguments, and identifying superficial, and not so superficial, errors.

PHILOSOPHICAL

INSTRUMENTS

1

SCIENCE, TECHNOLOGY, AND
PHILOSOPHY

[Man] can adapt himself somehow to anything his imagination can
cope with; but he cannot deal with Chaos. . . . Therefore[,] our most
important assets are always the symbols of our general *orientation* in
nature, on the earth, in society, and in what we are doing.
—Suzanne Langer, *Philosophy in a New Key: Study in
the Symbolism of Reason, Rite, and Art*

The apparatus that generated a revolution in seventeenth-century science
were known as philosophical instruments. Telescopes, compound microscopes, and thermometers were hailed for their capacities to reveal new vistas of knowledge, empowering experimenters with the ability to discover
truths about previously inaccessible particles or corpuscles. As the frontiers
for inquiry expanded to reveal wondrous and sometimes bizarre events, the
metaphysical speculations advanced by the Medieval Schoolmen were denigrated as philosophical fantasies. The champions of experimental inquiry
could see new realms of beings by properly operating these new machines.
It was a glorious time for natural philosophy, not only because of the truths
these instruments unveiled but also because of the philosophical ideas they
confirmed. The stock-in-trade for designers of philosophical instruments
of the seventeenth century was both ideas and tools.

Although not used to characterize contemporary devices, the notion of
a philosophical instrument is neither trivial nor anachronistic. The seventeenth-century idea can be extended to instruments of twentieth- or twenty-
first-century research. By the mid-twentieth century, the natural sciences
witnessed another revolution in instrumental technologies, generating entirely new methods of inquiry. The profound transformation in research
techniques from "wet chemistry," for example, to a chemistry driven by the
fingerprinting techniques of electronic instrumentation had a profound
effect on "pure research." The instrumentation revolution of the twentieth
century was waged in the offices, conference rooms, and laboratories of the
chemical industry, responsive to the needs of manufacturers, government

agencies, and military institutions (Morris 2002). For example, invention and refinement of the infrared spectrometer were driven by work in industrial companies, particularly in the synthetic rubber program and for petroleum refinement. In organic chemistry, the new detection methods of the century were driven by discoveries in physical chemistry of the ways in which molecules are held together, bonds are formed and broken, and reactions occur at the molecular level.

The use of sophisticated research technologies carries with it specific convictions about the standards for knowledge. The alleged privilege given to pure researchers over instrument makers was undermined. The very boundary between detectable realm and unruly territory changed with the instrumental revolution of the twentieth century. More than merely an assembly of metals, wires, and plastics, the new instruments are channels of philosophical ideas about knowledge acquisition, ideas that are commissioned for laboratory studies. The new technologies reinvigorate the old notion of a philosophical instrument. Based on conceptions of human skill, instrumental power, and nature's properties, our inquiry reveals that tools blend with ideas in the (philosophical) instruments of the twentieth century.

In using instrumental technologies, researchers are committed to certain modes of inquiry. Instrument designers take a stance on how knowledge of materials can be attained and how such knowledge can be conveyed to others. Strict adherence to precise principles, commonplace in the history of philosophy, is not necessary and may even paralyze attempts to express how knowledge is gained from material skill.

I invite the reader to explore in the following pages how the twentieth- and twenty-first-century revolution in instrumental technologies demands its own standards of knowledge-inquiry as conceived in design plans for instruments and realized in laboratory studies. Important philosophical commitments about know-how are revealed in the standards of the so-called professional arts. Such commitments are not reducible to discursive knowledge from the theoretical sciences. As experimenters use instruments to explore the microscopic realm, unfamiliar vistas open and familiar ones are closed. The technologies that have radically transformed twentieth-century laboratory research with stunning results are true philosophical instruments, enabling researchers to reveal reliable information about the world. Of course, we must not force-fit philosophical prescriptions on the design of such technologies. Since instrumental technologies lie at the core of all contemporary laboratory research, attention to engineering design can yield insight into the experimenter/world relationship and reveal important commitments about inquiry. Philosophical ideas about knowledge acquisition are exposed, not imposed, through the careful study of the engineering of instruments.

Insight into the experimenter's relationship to the microscopic realm is gained by exploring the designs of mediating technologies. A philosophy *from* instrumentation, in contrast to a philosophy *of* instrumentation, emerges from such designs. Underpinning the engineering of information-generating technologies are commitments about knowledge acquisition. The designer imagines how a segment of the environment could be changed, anticipating the results of his or her possible action on the material world. Instrumentation requires both manipulation and analysis: experimenters manipulate, agitate, and transform a segment of the microscopic world, then analyze, monitor, and distill the results. Design plans define the instrument as well as the standards for attaining reliable information.

ACQUIRING SKILLS

For philosophers working in the logical empiricist tradition, the validity of findings from laboratory apparatus presumably relies on the same kind of epistemic standards for acquiring empirical knowledge through the naked senses. The laboratory instruments of the seventeenth and eighteenth centuries, such as the compound microscope, telescope, and air pump, function as mere tools, intermediate devices that extend the range of the senses. A skilled use of tools serves merely to enhance, correct, or validate the sensory experiences of observers. The methods of inquiry are quasi-transparent, and potential contaminants that arise from biasing factors can be removed from inquiry. Although demonstrating technical skills in their construction, such devices offer no special reward for the philosophical study of science.

None of the familiar theories of knowledge from the logical empiricist tradition in philosophy can adequately explain how knowing is achieved through the skilled use of tools. With the stunning advances in apparatus, many scholars today place instrumental skill at the forefront of laboratory research, dismissing familiar assumptions of rationalist philosophies. Between experimenters and an assembly of atoms are manipulative technologies, properly used. The notion of techné, referring to processes of creating artifacts through the skillful use of material, must be recognized as an essential and privileged aspect of inquiry. Experimental research is more an activity to produce new states than an application of abstract theories to a laboratory setting. A laboratory technician becomes an agent who, like the ancient artisan, starts a process, removes obstacles, and frees natural possibility from concealment toward the production of an artifact. In so doing, the agent exploits the concealed capacities of nature, hidden from immediate detection by the naked senses but revealed indirectly from the products of such operations. Validating data requires using apparatus skillfully. Rather than a secondary aspect of research, subordinate to the logic of confirmation, the production of artifacts is essential to data acquisition. The fundamental relationship between experimenters and the scientific world is invaded by the pragmatics of technical efficacy (Queraltó 1999; Agazzi 1999). The history of the physical sciences can be written as a chronology of successive discoveries of measuring devices. Machines dominate the historical landscape of research in ways that cannot be explained by rationalist philosophies of science. The practical ability of knowing how to achieve certain goals is an irreducible aspect of scientific practice.

Michael Polanyi argues against the a priori distinction between "knowing how" and "knowing that" (1969). The tacit knowledge required for the skilled use of tools is drawn from both the concrete, practical realm of individual experience and the abstract realm of theoretical prediction. But at some point, the performer must turn away from the immediacy of sensory experience and attend to participation in an entire experimental system (pp. 126–28). In teaching novices how to use tools, experienced hands move back and forth from concrete individual experiences to general patterns of associations among human bodies, tools, and raw materials. The old distinction between pure science, as the discovery of a given reality, and technology, as the deliberate production of artifacts, collapses for contemporary research. Science discovers through the production of artifacts.

Excessive attention to scientific theories—their character, construction, and confirmation—has generated utopian fantasies about inquiry, as Ian Hacking famously argues (1983). Between the proper use of the term "electrons" and the efficient use of electrons is a complex network of mediating agents, both human and nonhuman. Theories come and go, but the causal properties of certain entities deployed for purposes of research remain stable. Real-world entities are those that function as manipulating tools in the investigation of other, more hypothetical elements of nature (p. 263). Engineering, and not theorizing, reveals how certain entities are exploited to intervene in a specimen's dynamic properties.

In writings that followed *Representing and Intervening,* Hacking broadens his perspective to address the cultural dimensions of laboratory research. Stability of science is found in the mutual adjustment of ideas, materials, and marks (1992b, p. 30). Ideas comprise the questions, background knowledge, topical hypotheses, and modeling of apparatus; materials include the substance to be studied, apparatus, detectors, and tools; marks are the uninterpreted inscriptions, data processing, data reduction, and interpretation. All phenomena occurring in the laboratory are crafted artificially from the mutually adjusted segments of material practice (p. 49). The reliability of findings rests on adjustment of theory, apparatus, data, and much more (Hacking 1992a, p. 49). Any disunifying factor, such as unexpected data, can be assimilated by making suitable revisions to the relevant practice, retaining the system's closure from the real world. One segment of research can be vindicated by the mutual adjustment of the system as a whole (Hacking 1992b, p. 56). Theories have no privileged status in the system.

In a series of richly detailed studies of instrumental technologies, Davis Baird applauds Hacking's commitment to the manipulative character of laboratory research. Baird privileges the material culture of researchers (2004). The hands-on knowledge needed to operate sophisticated apparatus cannot be explained by theories about nature or prescriptions for rational science (Baird and Faust 1990; Baird and Nordmann 1994). Baird examines the cyclotron (Baird and Faust 1990), the grating spectrograph (Baird 1991), the direct reading emission spectrometer (2000b), and various other spectrometers made by his father Walter's company, Baird Associates (Baird 1998). In some cases, standards for "objectivity" give greater weight to social and economic factors driving industrial production of the device than to accuracy and precision of the technique itself. Cost efficiency supersedes instrumental accuracy in the notion of objectivity of technique (2000a, p. 110). Chemical knowledge is judged more on research techniques associated with manipulating and controlling materials than on the theoretical representations of microscopic processes that presumably provide the rationale for such techniques (2002). A well-known manufacturer of analytical instruments, with cutting-edge contributions to emission and infrared spectroscopy in the 1930s and 1940s, Baird Associates illustrates how the instrumentation revolution changed forever how scientific knowledge is acquired and the way science and technology are mutually supportive.

Much debate over the unity/disunity of science centers on the role of agency. David Gooding provides a detailed study of the material, cultural, and (individual) human agents that determine experimental practice as he constructs a complex mapping system of a researcher's thoughts and actions in his analyses of the experiments of Michael Faraday (Gooding 1990) as well as of the quark-search experiments of Giacomo Morpurgo (Gooding 1992).[1]

Andrew Pickering examines how material and social resources are brought to bear on the daily work of laboratory physicists. The physicist maneuvers in a field of materials, constructing machines, putting segments of the environment in service as if domesticating nature (1995, p. 7). Artifacts of a scientific culture bring together human and nonhuman agents within networks of capacities (p. 21). The properties of material agents are manifested in a world of human agency, typically through the application of skills, under the pressures of cultural factors associated with the production of machines. Conversely, the network of human agency is revealed in performances associated with the production of artifacts. Ultimately, the practices are mangled from an inescapable tension between the need to accommodate some agents (material, social, culture) and the need to resist others (p. 23).

Through their sociological studies of science, Harry Collins and Steven Yearley attempt to redress the mistaken privilege given to science-as-theory doctrines. Experimental researchers are driven by the same kinds of cultural factors that determine the participants in our everyday social world (Collins and Yearley 1992). In her rich studies of biological laboratories, Karin Knorr-Cetina also argues that laboratory practices are deeply anchored in the culture of daily life, a life occurring outside of scientific research institutions (1992). A laboratory culture comprises various material, social, and conceptual elements, not amenable to a single systematic order. In his wonderfully detailed explorations of a physics laboratory, Peter Galison argues that a laboratory is a site for competing scientific cultures, that is, a trading zone for exchanging goods. Both sides impose constraints on the nature of the exchange without threatening the identity of either culture. Trading requires consensus about the procedure for exchange, determining which goods are valued equally (Galison 1997, p. 803). The interaction between competing scientific cultures requires local (but not global) coordination, without sacrificing the identity of either culture.

Pierre Laszlo examines another element of laboratory culture (1998). For Laszlo, the physical display of laboratory tools is rationalized by utopian pretensions about their purpose, masking their actual function as rhetorical devices. He defines instruments as inscription devices for constructing the texts of science. Rather than revealing the real-world properties of atoms and compounds, Laszlo argues, the new devices are used to convey prestige and power as chemists proudly show off their contraptions to visitors or advertise their techniques to journal referees. The so-called revolution in instrumentation enhances the mystique of researchers and perpetuates a professional myopia, as he puts it, about the rhetoric of chemistry.

Bruno Latour analyzes scientific practice through the metaphor of forming and breaking alliances. His analysis treats the human actors and natural agents in instrumental technologies similarly as "participants" in struggles for political power and social prestige. For Latour, scientific culture is not mangled; it is two-faced. One face reveals ready-made science, where scientists construct efficient machines, get facts straight, discover universal laws, and acquire theoretical knowledge. In ready-made science, disputes are settled because nature's properties are discovered. However, the other face reveals how science is unsettled in processes of making machines and selecting instruments for research. As long as controversies are rife, nature can never be used as a final arbiter (1987, p. 355). The laboratory apparatus, research facilities, and instrumental techniques therefore become rhetorical devices in local skirmishes, judged for their value in persuading potential critics. Latour believes that declarations of success have nothing to do with accessing nature's real structures. To secure victory and to silence opponents, the winners of a controversy strategically invoke the *idea* of nature and impose *rules* for proper research. But no epistemic basis for such pronouncements is possible, because no commitment to physical reality, existing independently of social and cultural influences, is warranted. Latour writes: "We can never use the outcome—Nature—to explain how and why a controversy has been settled" (p. 364). Any pronouncement about the discovery of the double helical structure of DNA, for example, merely serves the needs of a propagandist in a campaign for power.

Hacking's attempts to champion engineering as philosophically significant should be applauded. I agree with his important claim that experimental phenomena created from the use of modern instruments are inseparable from the localized setting, and the artifacts of an experiment are never reproducible outside of those same laboratory conditions. Experimental entities are real through a sound production process that is inseparable from many technological interventions, manipulations, transformations, and disruptions. Such information generated by experimental phenomena is not contaminated by technological inducements and not necessarily tainted by their inseparability from localized conditions of research.

But neither Hacking nor any other scholar cited above engages in a detailed reflection of design plans per se. Instrumental design offers insight into the ways in which experimenters acquire knowledge about remote regions of the world, and a critical reflection on design plans for instruments will reap philosophical reward. Critical reflection on design plans for instruments, which I offer in the following pages, reveals an interdependency between knowing *how* to engage in research and knowing *that* certain causal processes can be exploited. Yet, when facing questions about ontology of nature, we should avoid the temptation of privileging engineering at the expense of theoretical developments. Designers of instruments are often motivated by advances in the physical sciences and implicitly committed to the ontology of the relevant theories. The recognition of engineering as philosophically significant for a thorough understanding of scientific inquiry should not support the claim that scientific theories fail to reveal real-world causal processes. On the contrary, in engineering, causal models drawn from scientific theories function as cognitive tools, as it were, used for technological innovation, and such tools affirm the ontological commitments associated with the theories.

USING PHILOSOPHICAL INSTRUMENTS

For Suzanne Langer, some of the greatest human achievements are found in the symbols that govern general orientation in nature, on the earth, and in society (1942). Such symbols can take many forms: drawings, stories, principles, or theories. In indigenous societies, purification rituals provide orientation in a spiritual world. The ritualistic uses of water, fire, or food function as concrete embodiments of ideals, beliefs, and values associated with the cosmic order. The participant's presence here and now is located securely within that order, providing an effective remedy to the fear of a chaotic world.

Although we are not accustomed in the post-Enlightenment world to thinking of scientific instruments as akin to symbols, the skillful use of instruments relies on symbols of orientation, of finding one's place through a process of knowing. For example, a spear is not a microscope, but in many ways the differences between these artifacts are fewer than one might imagine. Like a microscope, a spear is an instrument for knowing one's place in the world. Our first toolmaking ancestors who fashioned lengths of wood, bone, and flint into weapons utilized more than mere animal strength and hand-eye coordination when they set out to hunt. The feel of the spear in the hand as it pierced an animal's hide, its effective (and ineffective) deployment against various prey, and a host of other forms of feedback made the hunter-spear unit an efficient and deadly combination. Put more simply, the world through the eyes of a spear-user was divided into opportunities and risks that emerged from tool usage, appearing as potential targets and impediments.

The skilled use of any tool enhances and reestablishes its user's particular orientation to the world in various contexts. For example, using a spear involved an understanding of how it as a weapon enhanced the hunter's range and potency of action—a spear could be thrown to increase the reach of the arm, and its sharp edge could pierce tough hides impervious to fists and small human teeth. Beyond the physical construction of the spear, however, the activity of spearing was perhaps primarily social, driven by the norms and rituals of the tribe (Ingold 1986, p. 6). Young hunters learned about spear construction, animal behavior, migratory and weather patterns, seasonal changes, hunting techniques, peculiarities of the locale, and much more from the wisdom of their elders. All of this knowledge was blended and refined through an ongoing catalog of physical experience that contributed to an ever-evolving corpus of hunting knowledge.

Today, carpenters, machinists, and specialists in crafts gain extensive knowledge about the behavior of materials through the skilled use of tools. With such knowledge, one's orientation to the physical world can be secured. As a novice learns familiar skills, or an old hand new skills, knowledge of past successes and failures with similar tools is distilled, communicated to less experienced hands, and applied to future endeavors. Such knowledge can be conveyed through verbal commands, written plans of action, and possibly documented guidelines. Any technological endeavor rests on foreknowledge of materials. Knowing *how* to change a portion of matter requires knowing *that* certain materials can be shaped, separated, combined, or transformed. Material skill demands a form of intelligence concerning an engagement that can be seen in the greater context of orienting with surroundings (Ingold 2000, chap. 16). Such foreknowledge involves awareness of those capacities of matter that can be put to good use. A master of any material skill is expected to adjust to surroundings that are unforeseen or not optimal for achieving a desired result. An understanding of strengths and weaknesses of materials was as pivotal for the earliest toolmakers as it is for contemporary designers of sophisticated technologies (Shapere 1991, 1999).

Whether we are considering ancient carvings in rhino horn or data inscriptions from the skilled use of absorption spectrometers, a material skill is definable through physical, cognitive, and social factors. First, in the physical realm, a skilled agent, or tool-user, exercises certain physical powers to change the material environment in ways that promote desired outcomes. Second, in the cognitive realm, the agent anticipates the effects of his or her manipulations, projecting forward to future courses of action based on a technical intelligence. To learn the skill of spearing, a young hunter had to understand both the strengths and weaknesses of the spear itself and the opportunities and dangers that lay ahead. The skilled use of a spear included some understanding of the material bodies, both in terms of their physical construction and use in the wild. In an act of cognitive projection, a skillful agent imagined how a segment of the hunting environment could be manipulated, anticipating the results of his or her possible action on the material world. Third, in the social realm, a skilled agent can monitor the results or test the product of an action based on a comparison to desired goals.

Today, a vast wealth of knowledge regarding material skill is captured in design plans for instruments. Manufacturers read such plans as prescriptions for construction, advertisers read them as marketing tools, and experimenters read them as signposts for research. One way for researchers to locate themselves in relation to microscopic processes, for example, is by reading design plans as ontological maps, that is, a mosaic of symbols for experimenters' orientation, marking out their place in the world through the skillful use of instruments. Of course, with any map, details are hidden, features are skewed, and landmarks are exaggerated. A design plan becomes a graphic depiction of possible explorations, identifying landmarks and paths for knowledge-seeking practices by identifying the opportunities for, and constraints upon, inquiry, as if dividing the experimental phenomena into "good" and "bad" events in relation to inquiry. In this respect, a design plan for laboratory technologies can be read as a channel of ideas about reality, dense with meaning about the idealized relationship between a skilled agent (experimenter) and a segment of the world (specimen). Design plans for laboratory technologies provide an idealized orientation between experimenter and specimen.

By bringing instrumentation to the fore, we reaffirm the unity of techné and ontology. Underlying the skillful production of artifacts are ontological commitments guiding researchers in procuring resources, both material and immaterial. Some of these commitments are discovered through engineering knowledge. Of course, in any instrumental technique, resources are deployed and manipulated for purposes of producing data. Material resources are strong/weak, good/poor, or efficient/inefficient with respect to their function in design. The well-worn theme that engineering is merely applied science must be abandoned because the physical sciences alone cannot capture the functionality of instrumentation.

A philosophical instrument can be defined as a means for knowledge-producing ends, enabling researchers to extend their reach to remote regions through the production of artifacts. Discovery of new techniques rests on overcoming distances between researcher and specimen based on an exploitation of known materials. In an attempt to compensate for the widening distance between experimenter and specimen, experimenters presuppose that the material of the world under investigation is endowed with the same kinds of capacities that designers attribute to their own constructed creations. In this context, a specimen exhibits properties of an artificial machine, manifested in its capacities to influence other bodies.

In this work, I take the stance that contemporary experimental technologies are philosophical instruments, broadly conceived. From this perspective, I divide the chapters as follows. In chapters 2, 3, and 4, I explore the claim that design plans consist of a mode of engineering knowledge, demanding attention to goals and strategies for constructing models of instruments. Chapters 5, 6, and 7 address the ontological commitments that underpin such models, revealed through a critical reading of design plans.

In chapter 2, I examine analogical modeling as a fundamental source of innovation in instrumentation. Of course, advances in the physical sciences frequently generate discovery of innovative research techniques. But the "pure" sciences alone cannot account for all aspects of innovation; a unique kind of discovery occurs when designers exploit analogies. In particular, designers of instruments privilege analogical models as a primary component of engineering knowledge, drawing upon processes from the natural environment for insight about the relevant causal mechanisms.

In chapter 3, I examine the use of computer-assisted design techniques. The graphic display of information through object-oriented programming languages has greatly improved the efficiency of the design process. Do such changes represent a fundamental transformation in design, with graphic images becoming a new and legitimate means for conveying information about a machine's structure and function? The manipulative power that such computerized techniques provide can be analogized to the capacities of an agent who dwells vicariously in a simulated environment through an active, practical engagement with bodies in a design space. I propose that certain doctrines of human ecology, well developed by social anthropologists, provide insight into computer-assisted design.

To what exactly does a design plan refer? The answer is not found in an actual material artifact, nor in any physical scale model of an artifact. As I argue in chapter 3, a design plan consists of an idealization, offering an abstract model of a hypothetical technique. A design plan invites vicarious participation in the machine's function, as a reader imagines how an instrument should be used, what changes will occur to the specimen, and whether desired results can be achieved.

Chapter 4 addresses the visual aspects of designing. In developing new laboratory techniques, designers resort to diagrammatical reasoning to visualize how a device will operate under experimental conditions. Through diagrammatic reasoning, designers establish an abstract space for simulating a machine's function. Visual modeling associated with conventional instrumental design is diagrammatic in a Peircian sense. For Peirce, diagrammatic reasoning demands three kinds of processes: (1) the introduction of new elements in the design space, (2) the interpretation of such elements as general concepts, and (3) the use of such concepts for hypothesis-construction and testing. I further argue that the visual skills of designers exhibit striking similarities to abilities of visual artists. The ability of artists to create perspective by representing edges, contours, cracks, and boundaries has always captivated viewers. The techniques for creating movement in works of art are important for many design plans for machinery. A reader experiences a kind of aesthetic movement in such plans by uniting actual perceptions of surfaces with an apprehension of possible features, envisioning a machine's actual surfaces and anticipating how possible surfaces would appear from different perspectives. In this way, the use of artistic skill to depict aesthetic movement creates the following perceptual problem: How can a schematic illustration convey both actual surfaces and possible surfaces of a material body? In some cases, the solution is achieved through the use of oblique projections, which are ubiquitous in both the visual arts and engineering design.

Chapter 5 examines the early improvements by Robert Hooke in the compound microscope, with attention given to his metaphysical commitment to nature's causal mechanisms. With a reputation as the foremost experimentalist in seventeenth-century England, Hooke nurtured his experimental skills in machine shops and in the glass industry of Holland. By learning the machinists' trade and improving on the production of lenses for his compound microscope, Hooke brought scientific inquiry closer to the Creator, revealing the universal principles of Nature of any material body, whether produced from the machine's tools or from God's hands. This chapter proposes that natural philosophers should learn the material crafts of research. The currency of these philosophical themes is explicated in light of discoveries made by analytical chemists deploying today's spectrometers.

In chapter 6, I examine the scanning tunneling microscope, reading in its design plans how knowledge is acquired through causal modeling. This technology illustrates the following generalization about every experimental instrument of analytical chemistry: the very determination of an analytical signal, generated through the use of an instrumental technology, rests on commitment to mechanisms of nature. The gap between laboratory researcher and detectable world is bridged through an indissoluble blend of apparatus and substance in the production of analytical signals. Of course, any signal carries information. But the pivotal distinction between desirable signals and noisy ones rests on a causal modeling, guided by the theoretical sciences. A desirable signal finds its causal source in the specimen's dynamical properties; a noisy signal finds its source somewhere else, such as in an external emission of radiation. The validity of data produced at the readout hangs in the balance.

As stated previously, when a human agent engages nature through the skilled use of tools, certain material properties are recovered and brought into use for the purposes of human action. Nature responds to skillful engagement by revealing its machinelike capacities to generate detectable events. A specimen's reactive capacities are exploited during instrumental manipulation, revealing fresh perspectives on its mechanical nature. In this respect, the machine metaphor retains its hold on contemporary researchers but is given new meaning from current technologies. As I argue in chapter 7, a specimen functions as a source of information during research, definable by its capacities to emit signals under appropriate technological inducements. With the proper technique, a specimen's causal capacities are manifested in the signals it produces. Nature functions as a causal machine, endowed with capacities to generate movement when sufficiently agitated, as Nancy Cartwright argues (1999). The causal mechanisms that drive the machine give rise to regular behavior that is conveyed through scientific laws. During research with instrumental technologies, a specimen exhibits machinelike capacities by functioning at times as an agent for change and at other times as a reagent responding to external influences.

SPANNING THE EXPERIMENTER/WORLD DIVISION

From a purely physicalist perspective, an instrument's function is reducible entirely to real-world causal processes that follow the models of the theoretical sciences. But a physicalist reading of design plans for instruments is incapable of capturing the rules, prescriptions, recommendations, and prohibitions for skillful practice and therefore unable to warn experimenters of opportunities and risks in the acquisition of information. In defining an instrumental technique, designers move back and forth between its technical function and its physical structure, attributing a dual significance to instrumentation. The technology associated with information processing emerges from causal models of the physical sciences, which in turn are derived from a distillation of empirical results from technologically induced manipulations of the past. In experimental research today, neither the physics of physical structures nor the engineering of the technology is privileged, and current research practices cannot support a reduction of physics to engineering nor engineering to physics. A complete understanding of the scanning tunneling microscope, for example, requires both functional and structural descriptions. The physics of quantum tunneling and the instrumental technique of scanning tunneling microscopy rely on the same physical ontology. Yet, in experimental research, the familiar distinction between natural object and nonnatural object loses its meaning (Kroes 2003, p. 85).

When philosophical instruments are deployed, the categories of human skill, instrumental power, and a specimen's capacities are appropriated for research. As laboratory events come and go, agency endures, lying in wait for another appearance. Underpinning each experiment are the capacities of apparatus, the abilities of experimenters, and the properties of raw materials. The design plans for experimental technologies uncover knowledge of both artificial materials and "natural" processes in the form of causal models from the physical sciences. With advances in laboratory technologies, the limits to experimental research continually recede, enlarging the scientific world, increasing the range of inquiry in both space and time, ever challenging our notions of common sense with ostensibly "bizarre" discoveries. As instrumental techniques improve, newly discovered objects seem increasingly distant from us: distant in space and in time, distant ultimately because of their strangeness with respect to the laws of the workaday, macroscopic world in which we live (Pomian 1998, p. 228). Engineers are continually faced with new challenges for discovering what is possible, drawing upon knowledge of what is actual.

Fundamental notions about inquiry function as cognitive resources in the design of laboratory instruments (as techniques). Twentieth- and twenty-first-century instruments (techniques) are laden with philosophical presuppositions, which are not expressible through a metalanguage of logical structures but can be read in plans for designers. Philosophical ideas about research emerge from a critical reflection on the constraints upon, and opportunities for, skillful practice.

I begin my exploration with a study of design plans for instruments. A design plan offers guidance for future practices through a visualization of the ideal instrument function. Just as chemists yearn to "see" the DNA molecule and astronomers try to "see" the galaxies, engineers construct design plans to visualize the opportunities and obstacles associated with a new device. A viewer imagines being transported into deeper regions of the empirical world, encountering remarkable experiences of scenes of minute detail. Via design plans, one acquires a general orientation to the portion of the world under investigation.

2

ANALOGIES OF DESIGN

Designers are responsible for anticipating possible problems that might arise for manufacturers and customers of instruments. For example, what if certain elements are subjected to external mechanical forces that cause displacement in space? How would the elements change internally as a result of a physical influence of heat? What unintended events might contaminate the results? For centuries, engineers have relied on knowledge of natural occurrences, moving back and forth between the external world and the world of instruments to address design problems. Natural phenomena function as exemplars for the properties of artificial materials. Eighteenth-century experimenters used scale models to simulate earthquakes, the formation of waterspouts and whirlwinds, the Northern Lights phenomena, and the behavior of light in general. In biomechanics, organic processes are often exploited for technological innovation. Designers are constantly charting a path to a promising future based on reports of past analogous designs. The assembled artifact will reveal similarities to known technologies, retaining their virtues while circumventing their weaknesses. In such cases, real-world processes are exploited for their analogical associations to artifacts. Obstacles to experimental detection are often resolved by identifying known processes, thereby exploiting analogies between familiar and unfamiliar events. The examples below offer clear evidence of the centrality of analogical modeling to instrumental design.

ANALOGICAL MODELING IN SCIENCE

Laboratory researchers often devise analogous experiments, especially in early stages of a research tradition. Major advances in experimental research are often achieved through analogical modeling, bringing insight from one field to bear on laboratory research. Advances in optics, for example, were often achieved through analogies to acoustics. In his *Opticks,* Sir Isaac Newton explained the light spectrum that he produced with a prism through properties of sounding bodies, such as bells. For Robert Hooke, the properties of musical strings provided insight into the universal character of all machines, both artificial and natural, as examined later in chapter 5.

The nineteenth-century biologists developing the new field of organic chemistry explicitly relied on analogical modeling for guidance in research. Reminiscent of the alchemists' attempts to imitate nature in a laboratory, researchers investigating organic phenomena drew upon findings from the "outside world" of inorganic chemistry. During the 1830s, the great Swedish chemist J. J. Berzelius modeled his research of organic compounds on the electrochemical principles of acids and bases in the inorganic domain. Inorganic salts are reducible to an acid, with its electronegative pole, and a base, with its electropositive pole (Brooke 1973, p. 77). As in the inorganic compounds, organic substances are neutral compounds containing two parts held together by some sort of electrical attraction. Berzelius advanced the following principle of analogy: Only from comparison with inorganic compounds can researchers develop ideas about the atomic grouping of elements or organic compounds (Brooke 1980, p. 42).[1]

The French chemist August Laurent also identified analogies between organic and inorganic phenomena, but his analogies went in the opposite direction (Brooke 1973, p. 64). Laurent revised research techniques for studies of inorganic compounds based on his findings in organic chemistry, reinterpreting concepts about inorganic elements in light of the structural complexity of organic radicals. Seeking to reduce organic compounds to the relative positions, arrangements, and order of constituent radicals, he redefined the notion of an organic radical, not as a group of elements that behaved like inorganic elements, but as a hydrocarbon prism itself, within which substitutions could occur (Brooke 1980, p. 44).[2]

The role of analogy in science has provoked lively debates among philosophers, historians, and scientists. The controversies center on three major questions.[3] First, what exactly is an analogy? In its broadest sense, analogy is a resemblance between any two things. Advocates of analogy direct our attention to the vast intellectual treasures that scientists produce from certain kinds of analogies. One of the first known uses of the term "analogy" appeared in mathematics of the ancient Greek thinkers, referring to a proportion between two numerical ratios. The term was extended to convey a resemblance, or shared pattern, between observed objects in the world. For example, advances in optics were achieved by exploring analogical connections between properties of light and those of sound. The origin of organic chemistry in the mid-nineteenth century was prompted, in part, by reinterpreting biological processes in terms of crystalline properties of minerals.

Second, what is the relationship between the units (relata) of an analogy? From a purely logical perspective, any one thing exhibits some resemblance to any other thing. According to Mary Hesse, the relationship is either formal or material. A formal analogy requires a binary relation between two types of objects X and Y, whatever the objects happen to be, and whether or not they are real. We say that the geometrical pattern of tiled floor in a bathroom is analogous to the two-dimensional pattern of cells of a honeycomb, since the tile and honeycomb cells exhibit the geometrical properties of a hexagon. On the basis of a formal analogy, certain properties (or elements) of one set of objects X are mapped onto those of another set Y, according to principles of mathematics. Beyond the requirements of a formal analogy, a material analogy establishes certain material similarities and differences between sets X and Y. Every material analogy exhibits known similarities (positive analogies), some known differences (negative analogies), and some neutral analogy. A neutral analogy conveys relations of unknown similarity and unknown difference and thus functions as a potential source of new ideas.

A third controversy surrounding scientific analogy centers on the following question: What is the role of analogy in the advancement of science? Critics argue that scientific analogies at best serve as mere didactic props and always produce a fictional parody of the real thing. But for advocates of analogy, insights about worldly events can be discovered by exploiting three types of analogical associations. (1) Analogy depicts the relationship between an observed pattern of events and a theoretical model constructed to replicate such a pattern. A conceptual model simulates such events by producing an analytical analogue to a segment of the world. If light rays are observed to "bend" when moving from air to water, a theoretical model offers a conceptual simulation of such a process. (2) The causal process that is responsible for light cannot be observed directly. To understand such a process, we need to exploit analogical associations between the causal mechanism for optical events and some hypothetical entities and forces with which we are familiar. For example, Newton understood light rays mechanically as the movement of imperceptible corpuscles, based on analogies to particles that move in Euclidean space. (3) The development of one theory is often advanced by its analogical juxtaposition to another theory. One theoretical model, called a donor, can function as a conceptual filter, or template, for redefining another theoretical model, called a target. The positive analogies between donor and target inspire scientists to explore the neutral analogies and to transform the meaning of the target.

The first generation of design methodologies in the early 1960s advanced problem-solving models of Herbert Simon (Dorst and Dijkhuis 1995, p. 261). Design was thought to be reducible to a search procedure, and permissible moves along such a sequence of states are given by operators, preferably in the form of known algorithms. Coherence of a design plan can be defined systematically, as illustrated by the celebrated systematic approach to design by G. Pahl and W. Beitz (1988). Problem-solving methodologies have been challenged on the grounds that they fail to account for ill-structured problems, contradictory constraints on the initial state and goal state, and unknown searching procedures. Algorithmic operations are rare in the actual design procedure (Goldschmidt 1997, p. 441). In ill-structured problems, the possible states are vague, incoherent, or simply unknown. The construction of analogical models is essential to successful design modeling. In many cases, a designer frames different views of the situation with respect to different perspectives, offering a range of possible solutions that could not be found by any single representation (McDonnell 1997, p. 458). Ill-structured problems require designers to conceptually re-represent the particular framework of an artifact by analogy to multiple models, exploiting various representation structures simultaneously (Oxman 1997, p. 333).

One striking case of analogical modeling is found in C. T. R. Wilson's design of the cloud chamber. As P. Galison and A. Assmus demonstrate, Wilson constructed the cloud chamber across two radically different research traditions: the tradition of nineteenth-century meteorology and that of Cavendish researchers engaged in atomic physics. In the late nineteenth century, meteorologists sought to explain natural phenomena by mimicking natural occurrences in the laboratory. The Mimeticists, as Galison and Assmus call them, devised techniques for imitating real phenomena of nature, such as cyclones, glaciers, and cloud formations. John Aitken, for example, designed a dust chamber for purposes of studying the dirty air of the nineteenth-century industrial cities of England. Aitken produced a miniature cyclone, just like the vast entity that could destroy towns and flood fields (Galison and Assmus 1989, p. 264). Wilson found considerable merit in mimetic techniques. During his studies on the summit of Ben Nevis in Scotland, he found that the atmospheric events "excited my interest and made me wish to imitate them in the laboratory" (Wilson 1965, p. 194). However, when Wilson directed his energies toward the actual reproduction of thunderstorms, coronae, and atmospheric electricity, he abandoned the Mimetic tradition in favor of the Cavendish style of analytic physics, reflecting the influence of his Cambridge education. Although Wilson was not engaged in ion physics per se, he moved freely back and forth between questions of ionic charge and the nature of atmospheric events (Galison and Assmus 1989, p. 257). In manipulating droplets in electric fields, he found no radical distinction between meteorological recreation and ionic experimentation (p. 804).

To Victorian meteorologists, the production of droplets around ions required miniaturization of the natural world of rain, fog, and thunderstorms. But unlike atmospheric events, laboratory phenomena are momentary states of a complex system, never isolated from apparatus, always produced from something constructed and something procured from outside. A phenomenon occurs when such an apparatus-world complex is sufficiently induced to generate artifacts. In his studies of atmospheric electricity,[4] Wilson created an artificial environment, an apparatus-world complex, by using air that was filtered through various devices, removing all traces of "dusty" air (Galison and Assmus 1989, p. 245). In a cloud chamber, moist air was expanded quickly and then supersaturated. If the expansion occurred beyond a specifiable limit, cloudlike features could be reproduced in the chamber. Ions carrying a known charge could be made visible by photographing the water droplets condensing along their trajectories within the chamber (Wilson 1965, p. 199). These streams of ions were measured by means of negatively charged plates, which were exposed to ultraviolet light in the cloud chamber. The track of an ionizing particle might be made visible and photographed by condensing water on the ions.[5]

MODELING MECHANISMS

To explain why certain events occur, experimenters often show that such events "came into being," as it were, through the workings of causal mechanisms. The analogical underpinnings of three particular instrumental technologies are presented immediately below.

Case 1: Spectroscopy was born from the need to explain the distribution of spectral lines. In the 1820s, scientists knew from the spectra of chemical flames that the rays of color indicate the presence of a definite chemical compound, but they could not explain the spectra. As long as the explanation defied comprehension, the use of spectral lines as a basis for qualitative information about the specimen remained suspect.

Spectroscopists of the nineteenth century were intrigued by analogies between atomic vibrations and acoustic vibrations. George Johnstone Stoney developed the original molecular theory of spectra by analogy with the laws for fundamental notes and overtones of a vibrating string and tried to explain how each spectral line corresponds to a specific movement of the emitting molecule. The solution to this problem can be found in the laws of harmonics: the wavelengths of the radiations emitted by vibrating molecules are, presumably, harmonically related (McGucken 1969, pp. 111–12). The reliance on acoustics as a prototype for the development of nineteenth-century chemistry was not entirely original, since analogies between optics and acoustics had been explored from the earliest days of microscopy. Hooke attributed a string's properties of length, tension, and thickness to all material bodies. For Stoney, the analogy of musical tones suggested decidedly antiatomic attributes: a medium has an internal capacity to vibrate. Only through its internal motion, which is undisturbed and regular, can a molecule produce light, according to Stoney.[6]

The commitment to a new prototype model in nineteenth-century spectroscopy did not incite a revolution in physical chemistry, as spectroscopists retained most of their cherished theories, methodological doctrines, and metaphysical beliefs about the nature of reality. Many of the central doctrines of physical chemistry at that time remained intact even as new prototypes were developed to overcome specific obstacles to spectral analysis.

Case 2: Absorption spectroscopy is an excellent technique for studying the transitions between quantum states. As an analytical instrument, these spectrometers are complex detection systems, transforming and processing information from an input event, typically an instrument/specimen interface, to some output event occurring at the readout. The electrical output is manipulated and sent to a readout device, such as a meter, computer, controlled video display, or printer/plotter. In its simplest form, absorption spectroscopy is a technique in which light shines through a sample, and the amount of energy that is absorbed, reflected, and transmitted through the sample is measured. That part of the radiation that passes through the sample is detected and converted to an electrical signal. The absorption of energy can sometimes alter the physical and chemical states of the molecules in ways that can be detected through a spectrometer. Although this technique never permits a direct observation of a molecule's properties, experimenters can acquire a great deal of information about the motion of the molecules, including the molecule's internal structure.

The success of spectral analysis is based on the following physical principle: If a specimen absorbs a certain wavelength of light, whereby each wavelength corresponds to a particular energy, then that absorbed energy must be exactly the same as the energy required for some specific internal change in the molecule or atom. Any of the remaining energies in the light spectrum that do not match that change are "ignored" by the substance, and these energies are then reflected or transmitted. The absorbed light energy causes such changes as atomic and molecular vibration, rotation, and electron excitation. As a result of this absorption, a specially designed instrument may detect an energy change, which may be "sensed" as heat, fluorescence, or color.

In contemporary research, absorption spectroscopy exploits analogies between energy impinging on the sample and photoelectric signals from the background models of electromagnetism (Rothbart and Slayden 1994). In order to absorb infrared radiation, which comprises radiations with wavelengths from 0.78 to 1000 um, a molecule must undergo a net change in its dipole moment. Only under these circumstances can the alternating electrical field of radiation interact with the molecule. In this context, the molecule is idealized as a system of point masses joined by vibrating bonds. The atoms fluctuate continuously as a result of different types of vibrations, similar to the simple harmonic motions of two masses connected by a string. In this context, a signal is explained through analogies to impulses of electromagnetic radiation from flashes of light. The analogical associations between signals and impulses of light invite scientists to extend various models of electromagnetism to a rationale for using absorption spectrometers.

Case 3: An abundance of biochemical studies revealing the conformational flexibility of DNA rely on Raman scattering techniques. In Raman spectroscopy, a sample is irradiated with a powerful laser source of visible or infrared monochromatic radiation, such as a helium-neon gas laser, which emits coherent, highly monochromatic light (Mann, Vickers, and Gulick 1974, p. 478). A small fraction of radiation is scattered by certain molecules, occurring when an inelastic collision between photons from the electromagnetic energy and a sample molecule produce a detectable transfer of energy. Under certain conditions, the molecule's energy state increases as it interacts with photons from an energy source. When the molecule returns to a lower energy level, the total energy of the photon is increased as the photon carries off the extra energy and its frequency is shifted up by some quantity, identified as the Raman shift (p. 475).

When Raman scattering occurs, the electronic field of a photon can deform, or polarize, the molecule's electronic cloud. Polarizability refers to the ease with which the electron distribution of the molecule may be altered, inducing a vibrational transition in the molecule. In such a case, the scattered light has been weakly modulated by at least some of the characteristic frequencies of the molecule. This occurs from a momentary distortion of the electrons distributed around a bond in a molecule, followed by reemission of the radiation as the bond returns to its ground state. This interaction between photons and the specimen composes a dynamic system in which the polarizability of the sample is changed (Strobel and Heineman 1989, pp. 633–34).

Raman developed his now-famous spectroscopic techniques by exploiting analogies to the Compton effect. By 1923, experimenters discovered that light can scatter in ways that could not be explained by classical electromagnetic theory. In particular, light that is scattered from collision with a molecule does not have the same wavelength as the incident radiation. The scattered radiation is diminished in intensity in order to reveal characteristic properties of the substance under investigation. The phenomenon, which had puzzled Raman and others since 1923, was in fact the optical analogue of the Compton effect (Raman 1965, p. 271). Raman extended a very important idea from Compton's research: that the wavelength of radiation could be degraded in the process of scattering. In this way, the scattered radiation appears with diminished frequency. Raman writes:

> In interpreting the observed phenomena [of molecular scattering of light], the analogy with the Compton effect was adopted as the guiding principle. The work of Compton had gained general acceptance for the idea that the scattering of radiation is a unitary process in which the conservation principles hold good. [So,] . . . if the scattering particle gains any energy during the encounter with the quantum, the latter is deprived of energy to the same extent, and accordingly appears after scattering as a radiation of diminished frequency. . . . [The] new method opened up an illimitable field of experimental research in the study of the structure of matter. (pp. 272–73)

For Raman, the principles underlying the Compton effect, such as conservation of energy, could be extended hypothetically to this otherwise puzzling phenomenon of molecular scattering.

In analytical chemistry, the rules for Raman spectra are not always applied under the same experimental conditions as are rules for infrared spectra.[7] This is explained by the difference in the basic mechanisms for each technique. Again, the mechanism for Raman scattering requires a change in the polarizability of the molecule. For infrared absorption, however, a vibrational mode of the molecule must change its dipole or charge distribution. Such a change is required in order for radiation of the same frequency to interact with the molecule and promote it to an excited state (Skoog and Leary 1992, pp. 299–300).

Raman activity can occur under conditions that differ from those needed for infrared activity. For example, carbon dioxide has four normal modes of vibration: two stretching vibrations are symmetric, and two other vibrations are asymmetric. A symmetric vibration causes no change in dipole, because the two oxygen atoms move simultaneously away from, or toward, the central carbon atom. Consequently, symmetric vibrations of carbon dioxide are infrared inactive. But this mode is Raman active because the polarizability fluctuates in phase with the vibration. In contrast to its symmetric vibrational mode, the asymmetric mode of a carbon dioxide molecule is infrared active, because the dipole moment of the molecule fluctuates in phase with the asymmetric vibration. In this case, the asymmetric stretching vibration is Raman inactive (Skoog and Leary 1992, p. 300).

ANALYZING ANALOGY

The cases mentioned above illustrate a general theme about engineering: that innovations in instrumental design rely on analogies for purposes of extending what is known about materials, phenomena, theories, and unobservable processes to what is unknown. In other words, so much of the engineering of instruments rests on exploiting metaphoric representations of natural systems.[8] Such modeling invites attention to multiple levels, from the observable similarities of materials to the conceptual similarities among multiple theoretical principles.

So, the first question about analogy mentioned above—What exactly is an analogy?—can be answered at different levels of abstraction, depending upon the context of the resemblance. Behavioral analogies rest on observable similarities between materials or types of phenomena. For example, the nineteenth-century Mimeticists observed similarities between natural phenomena and the dirty air of industrial cities and the miniature cyclones produced in dust chambers. Such behavior analogies prompted Wilson to extend, hypothetically, principles of ion physics to his design of a cloud chamber. He developed experimental techniques of ionic experimentation by exploiting known principles of meteorology, based on analogies between the atmospheric sciences and ion physics.

With respect to the second question above—What is the relationship between the units of an analogy?—we should recall that material analogies are fundamentally different from theoretical analogies. The behavioral analogies among observable material invite engineers to integrate theoretical findings from distinct disciplines. For example, in his development of scattering techniques, Raman extended by analogy the idea of the scattering of radiation associated with the Compton effect according to the conservation principles of physics. Raman relied on the Compton effect as a prototype for his innovation. In so doing, Raman retrieved certain physical principles in order to explain puzzling features of molecular scattering.

To address the third question—What is the role of analogy in the advancement of science?—we turn to the engineer's task of exploiting known strengths and weaknesses of materials for creating a technique that produces experimental phenomena for purposes of gaining information. Innovations in instrumental techniques will reward designers who can exploit the known to gain insight into the unknown, retrieving knowledge from the theoretical and engineering sciences to reveal new ways of manipulating and transforming materials in the service of research.

How exactly should a design plan be tested? One technique for testing design plans rests on a kind of thought experiment known as virtual witnessing, as I explore in chapter 3.

3

TESTING DESIGN PLANS

As discussed in chapter 2, advances in laboratory technologies draw heavily upon insights from analogical models, exploiting what is known to develop techniques for revealing what is unknown. In this chapter, I consider the following pivotal question about such models: To what exactly do instrumental design plans refer? We are tempted to answer this by appealing to the materiality of the instruments and then explaining properties of this materiality through knowledge of the theoretical sciences. But attempts to reduce instrumental design, and engineering in general, to the domain of the physical sciences are ill-founded. Although relying heavily on insights from the physical sciences, the design plan is both descriptive and prescriptive, replicating an instrument's structure and rationalizing its use during research. In their plans for instruments, designers do not directly refer to a tangible, material device. On the contrary, design, as a form of engineering knowledge, depicts idealizations, replicating *what* it is (the instrument) and *how* it should be executed (the instrumental technique). Designers of instruments are often responsible for explaining their work to manufacturers, justifying their research to funding agencies, or persuading reluctant experimenters to follow new techniques. Scrutiny of design plans by fellow researchers, prospective customers, and potential critics is constructive and typically requires visualization of a proposed technique.

In this chapter, I argue that design plans offer an idealized replication of inquiry that can be "tested" through a kind of thought experiment identified as virtual witnessing. We often read that thought experiments are confined to the realm of physics (J. R. Brown 1991, p. 31). But the design of an instrument can be tested through thought experiments without deploying metals, wires, and apparatus in a particular laboratory. Such testing occurs in a design space where certain processes are conceived and their effects anticipated. Designers invite critical appraisal of new instrumentation through the reader's imaginary participation in a hypothetical experiment, a thought experiment that requires visualizing the effects of instrumental manipulations.

A simple schematic illustration of an absorption spectrometer draws attention to an instrument's bulky (material) components (Parsons 1997, p. 262), as depicted in Figure 1.

Although this illustration is relatively superficial, it provides information about the functional relationships between components. Absorption spectroscopy is commonly used in chemistry for identification, structure elucidation, and quantification of organic and inorganic compounds. The material realm includes a radiation source, sample, monochromator, detector, and readout. A beam of electromagnetic radiation is emitted from a source and then passes through a monochromator. The monochromator isolates the radiation from a broad band of wavelengths to a continuous selection of narrow-band wavelengths. Radiation then impinges on the sample. Depending on the molecular structure of the sample, various wavelengths of radiation are absorbed, reflected, or transmitted. That part of the radiation that passes through the sample is detected and converted to an electrical signal, creating an event of the phenomenal realm.

FIGURE 1. Idealized illustration of an absorption spectrometer

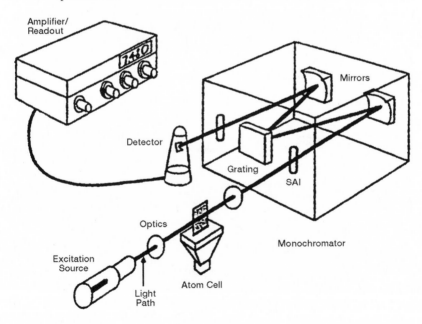

When an absorption spectrometer is used, photons from an artificial energy source bombard a sample, producing detectable reactions. Signals are produced, leading to inscriptions at the readout device. In this context, a source of energy that produces a narrow band of radiation is called a "line source" (Parsons 1997, p. 290). Designers of analytical instruments often use line segments to depict sequences of experimental phenomena. The term "line source" refers to a physical process associated with the progression of energy states. By reading the lines of Figure 2, one can understand how the beam path changes direction from contact with the instrument's components (Coates 1997, p. 442).

Of course, the lines depict the changing direction of the beam path from an energy source to the detector. Obviously, the surfaces of the material components, such as the detector, are grossly distorted and not true to scale.

In Figure 3, a radiation beam is conveyed by two shapes, not just a single line (Coates 1997, p. 445).

The pie-shaped segment depicts the progression of radiation from light source to holographic grating. The pie-shape exploits our perceptual skills of occlusion. The direction of the beam's motion is obvious, going from light to grating, and then from grating to array detector. The shape of the radiation beam that is projected from the light source can change with appropriate

FIGURE 2. Beam path of energy in absorption spectroscopy

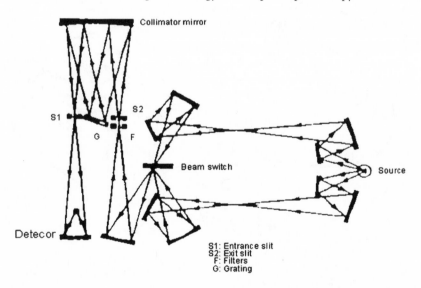

S1: Entrance slit
S2: Exit slit
F: Filters
G: Grating

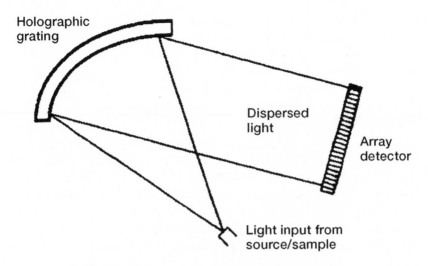

FIGURE 3. Pie-shaped depiction of energy path

changes in the energy source and with the shape of the grating. The shading from holographic grating to detector maps out another area.

Can the design plan offer manufacturers and experimenters assurance that similar (reliable) results can be attained under certain laboratory conditions? Has the designer anticipated most of the interfering factors that might contaminate the results? In this respect, reproducibility of results becomes a norm of experimental research, a hallmark of success. A design plan that cannot warrant trust in the reproducibility of results is a sure sign of failure. Of course, experimenters often discuss whether some empirical findings are reproducible. Talk of measuring dials, source of radiation, and performances of apparatus is replaced by talk of reproducible results. Researchers assume that certain aspects of an experiment are reproducible in other laboratories. An event can have its status raised, as it were, from something localized in a specific laboratory setting to a potentially recurring state.

However, some commentators argue that reproducibility is too demanding a standard because actual reproduction of an experiment rarely occurs. Financial, institutional, and technological pressures frequently impose insurmountable obstacles on the ability of scientists to reproduce an experiment. Occasionally, experiments are repeated when the researchers face stunning results or when the need arises to improve apparatus, augment data, or refine instrumental techniques. Typically, however, no one actually repeats an experiment (Hacking 1983, p. 231). After the results of a study are published in journals or disseminated through conferences, acceptance is achieved without reproducing the findings. This raises doubts about research practices: can the experimental findings of a particular study be validated without actually reproducing an experiment? If so, at what point in the process of validation does the community of scientists participate in the public hearing of these results? Can public participation be removed from the process of validation?

Hans Radder argues convincingly for the necessity of reproducibility as a criterion of success. He defines reproducibility through three categories (1996, chap. 2). First, we can speak about the reproducibility of the material of an experimenter, based on retaining material properties of an "original" experiment. In this case, the same material realization can be reproduced under different interpretations. Second, reproducibility applies to an experiment under a fixed theoretical interpretation when researchers resort to a theoretical description of an original experiment as a guide for an actual performance. A theoretical description is used to identify significant categories, distinctions, and relations in terms of repeatable properties. Third, reproducibility may also apply to the results of an experiment. The experimental results, q, remain stable while allowing for alternative procedures, p, p', p'', ..., pn, for realizing q (pp. 11–18).

Additional insight into reproducibility can be found in a process called virtual witnessing that S. Shapin and S. Schaffer explore in the context of Robert Boyle's experiments (Shapin and Schaffer 1985). Boyle needed to secure testimonies about the true results of his experiments and tried to perform experiments in a social space, a common practice in seventeenth- and eighteenth-century England. This became rather cumbersome because of the use of large equipment and of course could not reach a wide audience. He then commissioned engravers to create visual images of the experimental scene, such as the schematized line drawings used to imitate Boyle's experiments with the air pump. The viewer of such observable impressions was encouraged to generate mental images, as a kind of conceptual simulation of the experiment, in order to critically evaluate the experiment. Through virtual witnessing, readers could endorse the methodology and accept the findings without actually reproducing the experiment (pp. 60–62).

Returning to the contemporary scene, readers of design reports typically become virtual witnesses through their critical assessment of design plans. The reader rehearses, at least privately, the kind of malfunctions of apparatus, mistakes of implementation, and interfering effects that plagued past experiments. Are such dangers relevant to the present experiment? If so, are they avoidable? The answers to these questions require a command of the subject matter that is usually limited to expert witnesses. A reader imagines participating in an experiment for the purpose of evaluating the original plan of action. As a tool of persuasion, design plans provide readers with a cognitive vision of laboratory events, initiating a process of virtual witnessing. Readers are often persuaded that they *could* reproduce the same processes and would get the same correspondence of concepts to perceptions (Gooding 1990, p. 167). Without actually performing an experiment, a reader is expected to follow a narrative that selects and idealizes certain steps of a procedure (p. 205). This narrative transports the reader from the actual to the possible by "reenacting" the significant features of the experiment, focusing on the instrument's design, material apparatus, and microscopic phenomena, anticipating how a device would perform under experimental conditions.

Again, in defining an instrumental technique, designers rely on perceptual skills that are familiar in ordinary experiences of material bodies. Thought experiments produced from design plans include attempts to resolve problems associated with experiences of occlusion. The kind of visualization associated with instrumental design rests on graphic occlusion of edges and surfaces. In a design space, if we "move" around an occluding surface, previously hidden volumes become exposed and previously exposed ones become hidden. Also, movement through occluding surfaces will result in the exposure of previously hidden volumes at the expense of previously exposed volumes. As an experienced designer reads a nested series of illustrations from a particular design plan, information about embedded processes emerges. The reader engages in a thought experiment in which processes of increasing depth are virtually witnessed. The designer imagines acquiring enhanced powers of visualization to reveal processes, events, and structures of increasing depth. Models within models are revealed by imagining a change in relative position to visualize embedded processes. The thought experiment demands that a reader imagine being placed in a position to "observe" the unobservable. Through virtual witnessing, a reader imagines acquiring enhanced powers of visualization to detect embedded sequences of events, as if being transported along an axis of depth. New vistas appear as old ones pass by. Movement along an axis of depth presupposes a change in one's frame of reference.[1] When gaining information about experimental phenomena by unfolding layers of nested illustrations, a researcher exploits image schemata associated with an axis of depth.

By the end of the twentieth century, designers witnessed dramatic changes in modeling techniques, brought about largely by the burgeoning field of computer technologies. Molecular modeling has undergone revolutionary advances due to various computer graphic techniques. For example, the Virtual Reality Modeling Language developed in 1995 greatly enhanced the ability of chemists to replicate the structure of complex molecules with atomic elements and bonds (Mainzer 1999, p. 141). Using this language, designers can experiment on three-dimensional bodies in an object-oriented manner, modifying the position, orientation, and size of objects by interacting with a graphic representation. Objects can be grouped into more complex objects, and properties or operations assigned to the group are automatically inherited by its members (Emmerik 1991, p. 150). Various types of symbols (or nodes) are deployed in this language. Shape nodes depict points, lines, and shapes; property nodes illustrate color, texture, and geometrical transformations (Mainzer 1999). At the heart of these innovations are the automated computer programs for generating graphics, producing clear, vivid, and visualizable models. Computer simulation techniques engender a novel kind of experiment.

The engineering sciences have enjoyed great success with techniques of computer-assisted design, or CAD. Thoroughly integrated in design practices, such techniques are praised for their efficiency, speed, and low cost. The initial resistance to CAD-based modeling has been replaced by professional immersion.

Do such high-speed simulations alter the epistemic requirements of the sciences, demanding a fundamentally unique set of categories and principles of knowledge acquisition? According to one view, the obvious pragmatic virtues of smaller and cheaper machines, with faster executive times for modeling, do not necessitate a new epistemology. Although numerical analysis is black-boxed in the various computer graphic techniques, designers must still use the same mathematics and analyses to make sound decisions (Bertolline 1988).

But others find in computer simulations a radically innovative kind of experimentation necessitating a qualitatively different methodology in comparison to those associated with empiricist epistemologies of the physical sciences. According to this view, the traditional duality between theory and experiment cannot account for the primacy of computer graphics in the construction of visual models (Rohrlich 1991, p. 515; Petersen 2000). The reliability of computer-generated graphics is presumably a legitimate expression of knowledge, replacing the traditional priority given to linguistic-based symbols (Wright 1990). Such images are visual allegories of information, conveyed through analogies to perceptual qualities of a graphic scene in a design space. Seen in this context, scientific information is more like artistic imagery than abstract theoretical explanation (pp. 66–68). Relying on the mathematics of differential equations, computer simulations enable observers to compare images against images. Visualization has become the most effective means of identifying characteristic features out of complex dynamic data sets (Winsberg 1999, p. 290).

These arguments perpetuate a myth that visualizations are virtually absent from conventional design techniques, that image-based modes of knowledge are introduced through computer simulation techniques. Visualization practices in conventional modeling techniques are highly developed from centuries of research. The advancement of computer-graphic techniques continues a long-standing use of visual symbols as media for information, appearing in journals and texts. Again, the vocabulary of points, lines, and shapes has for centuries filled the journals and volumes of engineering. At various stages of engineering design, diagrammatic reasoning is required. The three elements of diagrammatic thinking are realized: the use of pictorial symbols, meanings given in the form of general concepts, and a reliance on a thought experiment as a means for resolving certain design problems.

The novelty of computer-graphic techniques in engineering is not explainable through the familiar epistemic categories of validating evidence, testing hypotheses, and confirming theories. Innovations of graphic techniques for modeling center on users' vicarious immersion in a virtual world. More than the virtual witnessing associated with conventional design techniques, computer-generated immersion requires participation in a simulated environment, subjecting "objects" to manipulating probes and then "observing" the results. Mapmaking is essential to such immersion. A location on a map is both a symbolic depiction of past experiences and a guide for travels to come, providing information about the opportunities and risks of interventions. Moving in a virtual environment is achieved by probing, exploring, manipulating, and looking around corners of "objects" in a design space. Information is retrieved through embodied participation in patterns of virtual events. Like navigators exploring unfamiliar territory, users gather information, situate themselves in a setting, and adjust to a course of action (Passini 1996).

The world's properties are detected when agents become immersed in an active, practical, and skillful engagement with constituents of an environment. For example, the environment is divided into favorable and unfavorable conditions for hunting, as the hunter learns how animals behave, where they are located, how they reproduce, how they develop kinship relations, and how to approach them (Ingold 2000, pp. 36–42). I offer an ecological interpretation of CAD. Immersion in a virtual environment is a metaphor for retrieving, distilling, organizing, and generally rendering information "rationally visible." Models are constructed to make sense of past action and to assist viewers in the engagement with objects of the virtual environment (wayfinding). The "objects," "events," and "processes" associated with computer-graphic modeling are not quite icons of visual models. They are rather surrogates for meshed patterns of possible actions, decisions, and mappings (Schubert, Friedman, and Regenbrecht 2001, p. 270).

Computer graphic techniques are employed as instruments for modeling. The so-called virtual environment is reducible to a set of rules and resources for modeling, defined by transformative relations among data. The term "virtual reality" is a metaphor for the rules and resources deployed as instruments for possible practices. Rules are deployed to identify constraints upon, and opportunities for, organizing data; resources are needed for production and reproduction, that is, means for an agent's movement through, and influence upon, the other agents in the environment. The "bodies" in a virtual environment simulate zones of past practice and provide a guide in future encounters. A virtual environment represents techniques for rendering information "rationally visible" to viewers. Descriptive modeling calls for the use of grids or zonal structures for ordering units of information. A virtual environment represents the cognitive instruments for maneuvering among such landmarks, instruments that establish transformative relations among data. Computer simulations are performed through a complex chain of transformations.[2]

In this respect, the grids provide rules for successfully navigating among the "bodies" of the simulated environment. The "objects" of the virtual environment function as geographical action-maps for permissible movement. For example, (1) the surfaces of a 3–D object in the virtual environment function as constraints on possible action; (2) the space between objects functions as an enabling condition, enticing possible movement; and (3) occluding lines function as boundaries between the actual surfaces and possible surfaces. Occlusion is particularly important for rendering data "rationally visible." An occluding surface functions as a visual metaphor for the distinction between information available to researchers at a given time and information that is potential, but not actually revealed, given through current detection methods.

A virtual world is formed through layered relations among categories and subcategories. As a viewer intervenes within a virtual world, surfaces are disclosed as others are hidden from view. Viewers gain a general orientation to the objects of the virtual environment by probing ever more deeply into their properties. Experiences with objects of a virtual environment entice the user to engage with objects at a deeper level, as if the computer graphics enable researchers to change the scales of resolution. With deeply layered models, stratification requires vicarious movement along an axis of depth, with scales of resolution changing through increasingly powerful probing techniques. Of course, such changes are metaphors for rules and resources for embedded modeling. Through stratified modeling, forms within forms are constructed as structures within structures are revealed.

One common result of utilizing computer graphic techniques is the stratification of analogical models. Immersion in the processes of a virtual environment is not captured by a single conceptual frame of data. Computer graphic techniques reward researchers with keen imaginative abilities to "see" multiple grids of information (Oxman 1997, p. 333). One system underpins the primary shapes of a virtual environment, and another system underpins the emergent shapes that experienced viewers can detect. Each new model introduces new modification strategies and operations for organizing data. The imaginative capacities of researchers are tested by the need to construct and manage multiple models from alternative reference frames. With conventional design methods, engineers visually represent the relationships between two variables through simple graphs or plots in printed form. But with new techniques, scientists can depict variables in a multidimensional design space and in so doing gain unprecedented freedom to monitor changes in a simulated environment. New vistas are exposed by projecting lines, shapes, and figures into the foreground. The degree of resolution can be changed, permitting a viewer to probe forms within forms, patterns nested within patterns. A viewer moves along an axis of depth. Surfaces are disclosed as others are hidden from view, enabling viewers to gain a general orientation to the simulated environment by eventually detecting bodies from many perspectives. Unlike conventional modeling techniques, computer graphic techniques enable researchers to change the scale of detection by magnifying segments of a simulated environment.

The following chapter addresses the creative aspects of engineering through a study of the icons and images of design plans.

4

ICONS OF DESIGN AND

IMAGES OF ART

"Well! I've often seen a cat without a grin," thought Alice, "but a grin
without a cat! It's the most curious thing I ever saw in all my life!"
—Lewis Carroll, *Alice's Adventures in Wonderland*

In the previous chapter, I underscored the importance of virtual witnessing as a
thought experiment for testing a design plan. Through this process, researchers
visualize the strengths and weaknesses of material, anticipate advantages and
disadvantages of the proposed technique, and imagine how familiar problems can be resolved and new ones possibly generated. The reliance on icons
of information is pivotal to virtual witnessing. In this chapter, I examine the
epistemic implications of icons as units of engineering knowledge. Not entirely
reducible to units of sensory data, icons provide designers with an efficient
means of conveying information about a machine's properties put into service
to express exactly what the machine is and how it will function. As a vehicle
for symbolic communication, an icon integrates known experiences with unknown but hypothetical forms. A decidedly creative aspect therefore emerges
in the selection and use of icons.

A diagrammatic mode of thinking also emerges from this inquiry. Diagrammatic thinking is manifested in the following aspects of design: first,
lines, shapes, and pictorial figures are introduced and manipulated in a design space; second, these geometrical symbols represent general concepts,
based on our perceptual skills of occlusion; and third, design plans induce
thought experiments that entice knowledgeable participants to evaluate the
designer's instructions to manufacturers, prescriptions to researchers, and
predictions to everyone about the prospects for success.

THINKING AND SEEING

The notion of a cognitive vision finds its place as a valuable topic in the history of philosophy. Aristotle identified intuition as a mental observation of

thought and error as visual illusion. The intellect is said to perceive an object through a metaphysical light. In this way, to know is to see. Vision and intellect establish a relation between the soul and the object looked upon (Pomian 1998, p. 211). René Descartes's insistence on clear and distinct ideas rested on his conception of intuition as the natural light of reason. Intellectual vision is an entry to metaphysics, providing knowledge of a body's primary qualities of extension and motion. For John Locke, intuition is a power to see relationships between ideas without the need for justification.

I appeal to a familiar theme from the history of philosophy: the application of general principles of reasoning to specific experiences rests on imagination to preselect concepts. Immanuel Kant, for example, argued that a specification of universal principles of cognition to experience proceeds artistically, not mechanically, in a temporal sequence of ideas. Not bound to the strictures of logic and mathematics, imagination is needed to subsume particular experiences under general concepts, requiring an inner sense of ideas (Makkreel 1990, p. 56).

Charles Peirce extended this tradition with the bold thesis that all rational thought is thoroughly blended with a kind of mental vision; all reasoning is a form of seeing. The mind forms visual diagrams in the imagination, producing skeletal patterns of segments of an environment. He writes, "By diagrammatic reasoning, I mean reasoning which constructs a diagram according to a precept expressed in general terms, performs experiments upon this diagram, notes their results, assures itself that similar experiments performed upon any diagram constructed according to the same precept would have the same results, and expresses this in general terms" (1976, 4:47–48). Through mental diagramming, the mind can represent a solution, which is then "tested" cognitively. Such tests are performed upon a diagrammatic representation and not through the use of laboratory apparatus.

For Peirce, diagrammatic reasoning requires (1) a hypothetical introduction of new elements not previously given in the definition of the problem, (2) the use of such elements as general concepts, and (3) the creation of new hypotheses for future tests. Such an experiment requires an active and imaginative search for physical circumstances for testing the proposed solutions. The solution can be tested vicariously through active experimentation in the imagination (Fernández 1993, pp. 236–37).

The analogical models of instrumental design are diagrammatic in a Peircian sense. Through diagrammatic reasoning, engineers establish a design space for simulating a machine's operation. A design space is an abstract representational space used to replicate a range of movements of components, based on principles of engineering and the physical sciences. Architects, artists, and mapmakers work in such spaces. Peirce's requirements for diagrammatic reasoning can be extended to the construction of design plans for instruments as follows: (1) new elements must be introduced hypothetically in a design space; (2) such elements must be used as general concepts for the detection of possible structures; and (3) new hypotheses are created for testing. A designer imagines how an instrument should be used, what changes will occur to the specimen, and whether desired results can be achieved.

Modeling in engineering is not fully explained through rational problem-solving models. Construction of design plans rests on diagrammatic reasoning, which requires a language of pictorial symbols, a semantics of general concepts, and a thought experiment of vicarious witnessing. In diagrammatic reasoning, certain nonverbal images, in an idealized form, convey meanings, giving structure to information in the form of a template or frame. Diagrammatic reasoning rests on organizing ideas through the use of visual structures, distilled from past experiences and used to frame forthcoming ideas. Images also are projective to unknown experiences. Explanatory power of the icon is found in its ability to make our experience of the world intelligible. In producing an icon of a possible reality, the imagination is not modeling something known but something that is in its inner nature unknown (Harré 1985, p. 32).

TOWARD A GRAMMAR OF PICTORIAL SYMBOLS

Belief that discursive language is privileged over visual symbols as an expression of empirical knowledge is almost universal in contemporary philosophical discussions. For logical empiricists, pictures, illustrations, schematic drawings, and diagrams serve, at best, as heuristic aids and, at worst, as sources of illusion when scientists need to convey information. The use of pictures is often denigrated, contrasted against the alleged superiority of mathematical equations, for example. To be sure, some scholars have recently challenged this conviction through rich historical and philosophical studies (Baigrie 1996). But the familiar dismissal of schematic illustrations in science remains a dominant position, tied to the conviction that scientific reason is expressible only in discursive form.

This conviction cannot account for the work of instrumental designers, who resort to pictographics to express functional relationships among components. For centuries, engineers have supplemented discursive descriptions with line drawings to show how an instrument will operate under various conditions. Illustrations rarely offer a cameralike mirror of the actual performance of a physically manifested machine and do not refer directly to this or that feature of a particular device. Experiments in contemporary research are often depicted through visual language, based on schematic drawings associated with an instrument's design. Chemical engineers use flow charts, electrical engineers use circuit diagrams, and many engineers use block diagrams as tools in design plans (Mitcham 1994, chap. 8).

In conventional design techniques, engineers convey information visually. Contemporary designer plans are constructed in various stages by deploying various types of drawings (Henderson 1999, p. 207). In early sketches, designers clarify tasks by specifying requirements of an artifact; project drawings first appear as early sketches that show proposals in broad outlines that conform to professional conventions and to the demands of various sources of funding; production drawings lay out all aspects of the material parts, providing dimensions and surface properties, establishing the material specifications, and determining the layout of components; presentation and maintenance drawings are developed after the production is finished for use by the customer; technical illustrations are used for popularizing books that use conventions of engineering drawing for public relations.

Visual literacy requires deciphering a code, as if translating a language. This is well known to engineers: "The knack of reading a schematic [illustration] is somewhat similar to, although very much simpler than, translating from a foreign language" (Mann, Vickers, and Gulick 1974, p. 45). Engineers are expected to comply with rules of meaning and are "reprimanded" if they do not. Pictorial images of unique shapes and lines are often included in a designer's vocabulary. A designer's vocabulary is a set of unique shapes, and a shape is a finite collection of lines. Design plans are often expressed through the arrangement of such shapes. To read visual information about instruments from such drawings, a literate observer must have some understanding of the grammar of pictorial symbols. Rules for a designer's language include a grammar of points, lines, and shapes. In this respect, the language of design is defined by a shape-grammar familiar to those working in a language of geometry.

A designer's vocabulary includes lines of various thicknesses, with different colors and intermediate cuts at different angles. Computer graphic plotters use pens with points of varying widths, ranging from 0.3 mm to 0.7 mm, for the purpose of drawing lines of different thicknesses. Figure 4 provides the example of "an alphabet of lines," according to one analysis of engineering graphics (Earle 1994, p. 187).

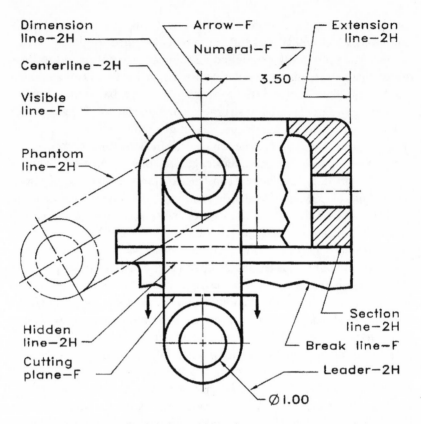

Dimension line—2H
Centerline—2H
Visible line—F
Phantom line—2H
Hidden line—2H
Cutting plane—F
Arrow—F
Numeral—F
3.50
Extension line—2H
Section line—2H
Break line—F
Leader—2H
⌀1.00

FIGURE 4. An alphabet of lines for drawing orthographic views

Information is conveyed in the foreground by the dashed edges and pencil thickness. In general, schematic illustrations are not drawn to scale, and the geometrical relationships between components are grossly distorted.

The symbols are used for conveying aspects of a machine's anticipated operation. Rather than corresponding directly to any particular machine, pictorial symbols convey information about abstract properties that define a mechanical system. Consider, for example, a table from a technical handbook on mechanical engineering showing some of the basic elements of a mechanical system, characterized by masses, springs, and forces. Table 1 introduces concepts associated with these elements and their impedances (Ungar 1996, p. 35).

For example, mass is depicted graphically through its relationship to ve-

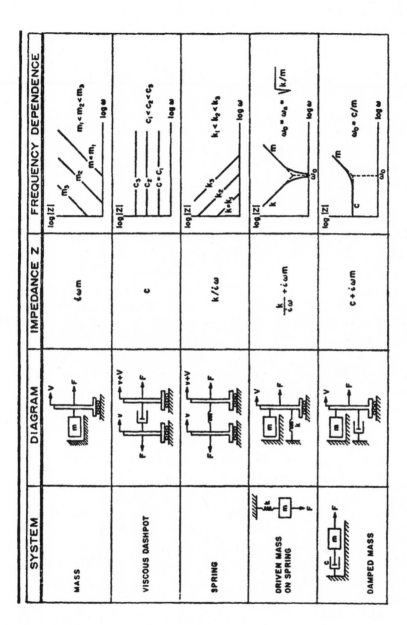

TABLE 1. Mechanical Impedances of Simple Systems

locity and force. The table functions as a kind of translation manual for the proper use of an engineer's pictorial symbols. Of course, such information must conform to theoretical principles of the physical sciences as well as to operational principles of engineering.[1]

The grammatical rules for any language are guided by practice. We follow a grammatical rule by engaging in our surroundings, just as we use an instrument by picking out and reacting to certain salient features of the environment. The visual character of language-use has affinities to visualizing how to manipulate bodies through the use of tools. A grammar is not completely free-floating in relation to worldly events. The commitment to a rule is inseparable from the ability to follow that rule, which in turn rests on how we experience a world. Rule-compliance requires skills for identifying known regularities and the ability to perform the same kinds of actions under the "appropriate" conditions. Certain kinds of experiences are salient and thus call for specific responses. Of course, a speaker's rule offers no mirror of reality. But a speaker follows a rule by visualizing how to *engage* a portion of our environment through certain kinds of manipulations. By reflecting on the grammar of schematic illustrations familiar to engineers, we can identify important aspects of an experiment. The shape-grammar in engineering design provides rules for the proper use of pictorial symbols. A shape-grammar exploits past successes and failures with similar devices and prescribes how experimenters should engage nature through proper use of the apparatus. The design plan places significant constraints on the proper means for investigating the world. By identifying the shape-grammar of a design plan for instruments, we reveal idealized standards for proper experimental inquiry.

VISUALIZING DESIGN

Engineering language includes a hieroglyphics of pictorial symbols. Visual literacy is required to decipher these symbols. Information about material bodies can be distilled, collected, saved, stored, put away, and then retrieved. In an engineer's design plans, lines appear frozen and unmoving, visually coded to depict invariant features in a sea of change. But certain pictorial symbols serve as interrogatives, inviting inspection at a deeper level. When reading design sketches, an experienced engineer reads more than visual cues on a "flat" surface. A design plan frequently includes a series of drawings, which are general at first but become progressively more detailed. In a design plan, lines of one schematic illustration gain specificity by association with other drawings representing deeper processes. When walking through a design plan page by page, it is as if a reader unfolds layers of information of increasing complexity. In the design plans for sophisticated instruments, information about embedded processes is conveyed in layers. A reader explores models within models, representing structures within structures. Such a sequence of models presupposes the lawful character of embeddedness. A reader assumes that smaller things (processes, events, or structures) are contained within larger things; events, occurrences, and processes at one level of depth are causally generated by mechanisms at a lower level.

Models generate a hierarchy of categories through the use of iconic symbols. In many cases, the symbols exploit certain properties of the system under development; the symbols themselves share certain features with the system. The visualizations associated with a design plan show striking similarities to the everyday experiences with occlusion. The experimental psychologist James J. Gibson discovered an important clue for understanding how observers can discriminate between different sources of disappearance. He examined how infants learn that toys and parents have a continual existence, even though they temporarily go out of sight. Surfaces disappear when the light is turned off, when another object blocks the line of sight, or when an observer moves his or her head in certain ways. Sometimes, when one surface disappears, other surfaces are exposed. Gibson demonstrated empirically that surfaces are perceived through the experience of occluding edges. An occluding edge has a double life: it hides some surfaces and exposes others. One's perception of an exposed surface is conjoined with awareness of hidden ones, linking possibilities with actualities.

A vertical line by itself is not occluding to an observer, because no information is provided about hidden surfaces. But Figure 5 depicts two figures with occluding edges, as the left figure shows two straight occluding edges, and the right one shows a rounded occluding edge (Gibson 1986, p. 81).

For each figure, a surface that is hidden in relation to an observer's line of vision could become exposed from certain movements. A realization of such exposure is included in an experience of occlusion. What would be exposed from an observer's movement to the left or right? An observer tends to "fill in the space" in ways that produce perceptions of a surface. Clearly, some aspects of the scene would change, such as the angles between line segments.

Occlusion is not limited to the experience of an exposed surface but invites attention to a realm of possible, but hidden, surfaces. An illustration of an apple's contours reveals a range of possible shapes and outlines, enticing an observer to envision how certain lines can be extended around corners and how surfaces can appear from different perspectives. Our fascination with

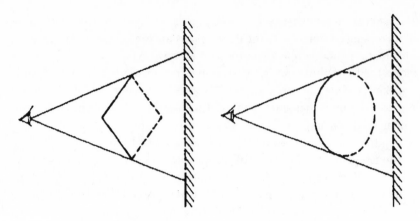

FIGURE 5. Edges of shapes are occluding in relation to a viewer's
line of sight

geographical maps is not explained by the triviality that maps reveal actual
regions. This value of maps can be understood through the perception of
occluding lines. Drawn by the lure of the next frontier, we rely on maps to
peer over that line dividing the known and the unknown, raising curiosity
for virtual travel.

An engineer relies on lines to depict features about a machine's operations. To an untrained observer, schematic illustrations seem static, providing a still-life vision of a bulky device. However, literate viewers can apprehend movement from such diagrams. An attentive observer anticipates how
our perceptions would change from the possible movement of objects that
clutter an environment. The lines of an engineer's design plan lie between
exposed surfaces and hidden ones in relation to a reader's line of sight. An
experience of occlusion for a particular material object demonstrates how
hidden surfaces can be exposed and revealed ones concealed by a change in
perspective. Whatever goes out of sight can return by changing the layout of
the environment (Gibson 1986, p. 92). Such awareness is enhanced by depth
perception. To apprehend contours of a house, edges of a fence, and outlines
of poles, we pick out certain features of the environment that remain invariant from one scene to another.

We perceive aspects of a scene by anticipating the effects of relative displacement. Surfaces that are momentarily blocked in relation to one's line
of sight can be exposed as a result of one's movement to the left or right, up
or down. In ordinary experiences of middle-sized objects, movement is reversible. Reversibility is an underlying requirement of our perceptions of an

occluding edge, a theme that Gibson articulates in his principle of reversible occlusion: "Any movement of a point of observation that hides previously unhidden surfaces has an opposite movement that reveals them" (1986, p. 193). So, some surfaces that are temporarily out of sight can be recovered.

In addition to identifying occluding edges, occlusion can also be attributed to surfaces. A surface can be occluding in relation to exposed and hidden volumes. In Figure 6, a surface depicting a wall of a house reveals one region inside and another region outside the house in relation to one's line of sight (Gibson 1986, p. 288).

In this figure, an observer can identify those surfaces that are located between perceived regions and those regions that could be perceived as a result of changing our reference frame. The hidden region is not completely unknown to a skilled viewer. In fact, in some pictures, more geometrical information can be acquired about the hidden volumes than about the exposed ones. In Figure 6, information about the shape of the exposed volume of a building's interior can be gained by an experience of occluding surfaces.

LINES, DESIGNS, AND CREATIVE MINDS

The rational problem-solving methodologies of the 1960s are too impoverished to explain the asymmetry associated with multiple projection systems. Iconic modeling in engineering design is often driven by spatial metaphors, as well as by image schemata drawn from embodied experiences and projected

FIGURE 6. Occluding lines depicting edges of a house

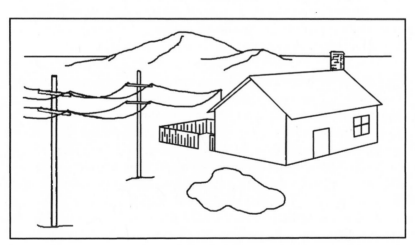

onto the target (embedded) system. Underpinning the presence of image schemas is a process of metaphoric mapping. Visual metaphors offer a creative response to design problems, especially valuable for generating new concepts, where familiar concepts from one context must be represented, transformed, and abstracted until they "make sense" in a new context (Goldschmidt 1997, p. 453).[2] As Mark Johnson argues, we organize everyday experiences with material bodies around certain images, such as up-down, front-back, part-whole, center-periphery (M. Johnson 1987, p. 267). These imaginative frames can determine what is "rationally visible" to others. Rather than being superimposed on data after the fact, an image schemata is thoroughly integrated in the relational ordering of information. Based on connections to patterns of past experiences, an image schema remains stable under the change of adding or removing information. When reading design sketches, for example, architects read more than visual cues on a "flat" surface. Architects have collected extensive data about the ways people talk about space. Landmarks are point-references external to the person, typically found in distant buildings or geographical features of a terrain. Paths are channels along which people move, such as streets, walkways, transit lines, and canals. Edges are linear elements along which two regions are related and joined (Miller and Johnson-Laird 1976, p. 375).

Visual tension in an artwork enhances awareness of one's own reference frame (Schummer 2003). Through productive ambiguity, a viewer becomes aware of his or her own presuppositions that determine experience. For example, in Renaissance perspective paintings, geometrical information depends on knowing one's position in relation to objects of a scene. An observer focuses on the invariant structure of light that underlies the changing perspective caused by his or her movements and projects geometrical axes onto a picture plane in ways that can be extended to other geometrical figures.

To develop these themes, I consider the following question: How can we explain the act of drawing a line? According to sense data theorists, line drawings provide a copy of one's sensory perceptions, as if the lines represent memory traces of past experience. A mental photograph is produced from a viewer's awareness of earlier perceptions of a visual landscape. Undermining claims of sense data theorists, artists have thoroughly discredited the notion that line drawings provide a copy, or photograph, of one's sensory perceptions. Even the term "copy" is a source of great confusion in the context of sense data theory. Other *drawings* can be copied, traced, and reproduced. In producing a work of art, an artist's mental images are never copied, but they can be replicated on a canvas.

A three-dimensional landscape that is flattened down on a picture plane scarcely retains the actual forms of the things depicted. Distortions are inevitable, because a picture plane is different from the surfaces that it represents. In three dimensions, two flat surfaces can be conjoined to form an edge or a corner; in a picture plane, the intersection of two planes is represented as a line. Lines in a work of art can be read as the contour to many possible surfaces, depending upon the ways that optical rays diverge, converge, or run parallel to each other. In Renaissance perspective paintings, vertical lines were used to convey a sense of alertness and reverence, as evident in the depiction of a cathedral; the horizontal lines of a landscape painting indicated stability; diagonal lines, used for action, indicated instability.[3]

Lines are significant not only to define shapes but also as cues for unseen shapes that lie outside of the immediate field of vision. A visual experience of a boundary invites attention to unseen, possible subshapes. As I argue above, the experience of boundaries is explained through occlusion: a viewer sees actual surfaces that are present in a scene and anticipates possible surfaces that might appear from different frames of reference. As the gaze moves around, under, or over an edge, some exposed surfaces become hidden, and previously hidden ones become exposed. We learn that solid bodies do not always disappear because of our own movement or because of displacement of the bodies in an environment. Some changes in perception have as their source the moving bodies of a landscape, and other changes are caused by an observer's own movement (Gibson 1986, p. 73). An occluding edge has a double life, placed between an exposed surface and a hidden one in relation to an observer's line of sight.

Occlusion is essential to visual modeling in design. The sensory skills of an experienced architect include the projection of subshapes that are imagined to lie behind, underneath, or within the manifested shape. The act of filling in shapes is a normal aspect of reading sketches, diagrams, and visual models of a design plan. The response to "missing information" is often a cognitive projection of subshapes and assumption of topological relations among subshapes. Through a shape occlusion, as it were, designers typically have a heightened ability to "see" emerging shapes that exist only implicitly in the primary, visible shape. Seeing shapes in an ill-structured problem often requires projecting subshapes and in so doing reconfiguring given shapes in terms of those emergent subshapes (Yu-tung 1995, p. 368).

An emergent subshape is a portion of a primary shape that is either closed or unclosed and explicit or implicit. From Figure 7, consider the following classifications of emergent subshapes in design: explicit closed, explicit unclosed, implicit closed, and implicit unclosed (Yu-tung 1995).

The four categories of subshapes can be defined as follows (pp. 375–76):

- Explicit and closed: shapes are both closed and located in front of segments of a primary shape (Figure a)
- Explicit and unclosed: shapes are unclosed and located in front of a segment of a primary shape (Figure b)
- Implicit and closed: shapes are closed but not necessarily located in front of segments of a primary shape. Subshapes can emerge by extending lines, connecting vertices, or connecting midpoints of lines. (Figure c)
- Implicit and unclosed: shapes are unclosed but not necessarily located in front of segments of a primary shape (Figure d)

Experienced architects, for example, have developed skills for projecting

FIGURE 7. Classification of emergent subshapes in design

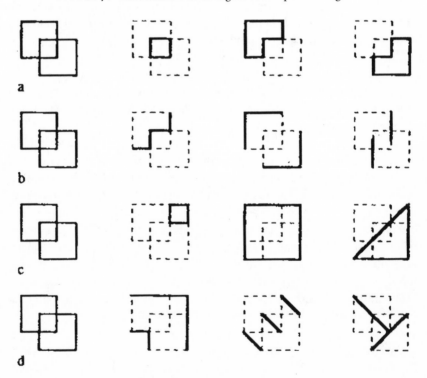

subshapes that might lie behind, underneath, and within a primary "seen" shape (Oxman 1997, p. 339). In empirical studies, researchers asked both experienced designers and inexperienced ones to identify the subshapes that emerge from certain primary shapes as well as the topological relations among them. Experienced designers "recovered" more emergent shapes than inexperienced designers (Yu-tung 1995, p. 382).

The recovery of emergent shapes from a design sketch requires skills that are well known in the visual arts. Again, the map of the visible intersects with a realm of possible images, as if the viewer experiences a transition from observation of actual given shapes to the recovery of possible shapes. Schematic illustrations of a design plan can be treated as works of art with a purpose. More than learning by doing, designers are constructing by immersing themselves in a design space, drawing upon their perceptual and cognitive skills to reveal features of "objects" in such a space. To decode the pictorial symbols of a design plan, a viewer resorts to skills familiar to visual artists. Such skills rest on a kind of aesthetic movement that artists have been attempting to capture for centuries.

Visual movement has been for centuries a major factor in the evaluation of artwork, long sought by artists, and prized by critics. This should not be surprising: artistic techniques are deployed to challenge and enhance a viewer's orientation to a scene. Artists know how visual experience unites visual images on a canvas with a viewer's possible perceptions. The artwork generates a tension between actual shapes and possible shapes, between a viewer's present experiences and those that would occur from changes in relative position. A viewer anticipates perceptions that would occur from various reference frames.

If movement is absent, the artwork is dead (Arnheim 1964, p. 341), for the greatest grace and life of a picture is expressed in motion, which painters call the spirit of a picture (Lamazzo qtd. in Arnheim 1964, p. 342). In the visual arts, the movement from actual to possible is a transition from exposed surfaces to hidden ones in relation to a viewer's line of sight. In his *Point and Line to Plane*, Wassily Kandinsky defines movement in line drawings as a tension in a certain direction (1979, p. 58). When deploying visual skills, a viewer merges actual perceptions with an apprehension of possibilities. Any work of art creates a visual tension, that is, a juxtaposition of actual and present experiences against the nonactual and nonpresent ones. Painting, like dance, uses aesthetic movement to create certain effects. The difference between the two art forms centers on the temporal sequence of perceptions. In dance, the order of our perceptions is prescribed by the work itself as part of the movement of the composition. In painting, the movement is not explicitly projected from the work but demands a viewer's active participation as part of the aesthetic experience (Arnheim 1964, p. 307).

Perceived lines in a painting are extended beyond the exposed surfaces of bodies in a particular "scene." The nonpresent and possible are tied to the present and actual work. If we "let a line go line," following Paul Klee's prescription (qtd. in Merleau-Ponty 1993, p. 143), then exposed surfaces merge with hidden ones. Our perceptions of enclosed shapes on the picture plane reveal one surface and conceal others. An artist is not producing a fictional version of the real aspects of an environment but is creating an imaginary texture of the real (Taminiaux 1993, pp. 284–89).

How can visual movement be explained? According to many artists, sensory perceptions of three-dimensional bodies are supplemented with a vision of features in a fourth spatial dimension.[4] The fourth dimension captures the realm of possible points, lines, and shapes, exposing otherwise hidden surfaces. One source of such movement in painting comes from perceptions of "emptiness." By outlining the edges of a solid body, artists can both depict surfaces of an object in positive (that is, a protruding) space and empty regions between the working components of the object in a negative (that is, receding) space. In his "Studies of Expressions for *Battle of Anghiari*," Leonardo da Vinci masterfully enhances movement by outlining the surface of the forehead (Baroni 1956, p. 133).

In Figure 8, a positive space between certain lines is "filled in" by a viewer's imagination, and regions of negative space are revealed where no surface of a material body is outlined. Information about surfaces can be conveyed by an absence of any image; an awareness of absence becomes a property of presence. Of course, this effect is enhanced by shading around the contours of the figure, mimicking a light source from above. As Alberto Giacometti writes on the topic of negative space: "I saw the emptiness" (qtd. in Arnheim 1964, p. 89).

Of course, depth perception in works of art can also be enhanced through certain techniques. An experience of depth perception is enhanced through painting. To add a third dimension to a flat visual field, artists often resort to linear perspective and the use of light and shade. The impression of depth is also created by the supposition of figures on a picture plane. Certain pictorial symbols are seen as overlapping, revealing surfaces of distinct material bodies. The detection of multiple surfaces can represent a dimension of directed action, which is another instance of aesthetic movement (Arnheim 1964, p. 343).

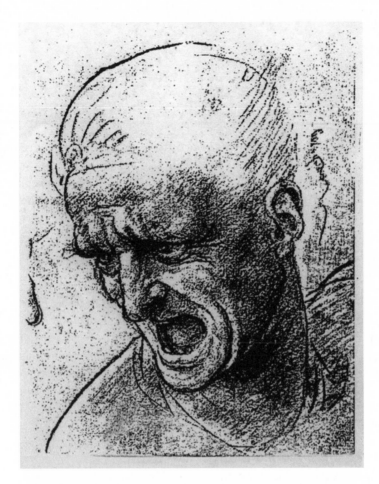

FIGURE 8. Sketch from Leonardo's "Studies of Expressions for *Battle of Anghiari*"

Aesthetic movement is enhanced by the use of certain geometrical shapes. For example, *a, b,* and *c* in Figure 9 clearly convey overlapping figures, as if one figure obstructs a view of another figure in relation to a viewer's line of sight (Arnheim 1964, p. 200, figure 186).

The eye of the viewer glances from front to back, and vice versa, producing an effect of aesthetic movement and imaginatively establishing the objects in a three-dimensional space.

Aesthetic movement can be explained as a second-order transition in the matrix of relations among shapes. Such a matrix renders the visual inscrip-

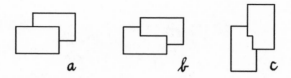

FIGURE 9. Aesthetic tension generated by overlapping shapes

tions on the artwork coherent through the use of a particular projection system. Through aesthetic movement, one imagines the juxtaposition of different systems of ordered relations, one system undergirding the primary shapes given in the work and another ordered system undergirding the emergent shapes that experienced viewers can detect.

THE EPISTEMOLOGY OF ORTHOGRAPHIC PROJECTION

Some artists exhibit tension in their work by developing an interplay among multiple projection systems. A projection system offers a visual perspective in terms of the intersection of rays of light from the scene to the picture plane. In representational art, a typology of a projection system begins with the distinction between orthographic projection and oblique projection. In orthographic projection, surfaces of objects intersect the picture plane at right angles in both directions. The picture plane shows the front, side, or top faces of material bodies, as if the object had been "rolled up" with printing ink and then pressed down on paper (Dubery and Willats 1983, p. 14). Orthographic projection underlies the design plans of engineers and architects, offering the "true" shapes of material objects.

In contrast to orthographic projections, the oblique projection has its rays intersecting at an oblique angle, as if the shapes are "drawn out" in one direction or another. In horizontal oblique projection, the shapes are drawn out horizontally because the projection rays intersect a picture plane at an oblique angle in the horizontal plane. In vertical oblique projection, the projection rays intersect the picture plane at an oblique angle in the vertical plane. A single projection system used to identify shapes and relations among shapes is combined with another (or many) projection system(s) used to identify shapes and the relations. So, the perception skills required to identify this or that shape draw the viewer's attention to the projection system needed to define a shape and to identify relations among shapes. Aesthetic movement that a viewer experiences in an artwork is rendered coherent as a cognitive change of reference frame, exploiting abilities to perceive the artwork through multiple projection systems.

For centuries, engineers have conveyed geometrical information through orthographic projection. To specify the height, width, and depth of an artifact, a designer often shows, in separate illustrations, the front, side, or top faces of an artifact. The front view conveys only width and height. The top view is projected onto another projection plane, and the side view is projected onto a third projection plane (Earle 1994, pp. 185–86).

In representational art, orthographic design produces a frontal orientation of flat surfaces, as if the face of the object were pressed down on the picture plane. Optical rays from the surfaces of a material body to a picture plane are parallel to each other and intersect the picture plane at right angles. In Egyptian art, orthogonal projections convey parallelism, perpendicularity, and relative lengths and areas (Hagen 1986, p. 187).[5] Of course, no single picture conveys the third dimension. Orthogonal projection presumably overcomes the problem of depicting the three-dimensional shape of objects by providing multiple illustrations of different faces (Willats 1997, p. 44).

However, orthographic projections alone provide isolated snapshots of surface features in a series of static images. By inhibiting aesthetic movement around, under, or over material bodies, the tension between the actual and the possible, so important for graphic occlusion, cannot be easily experienced through orthographic projection. The so-called fourth dimension of a design space is difficult to retrieve from such static projections. Consequently, orthographic projection tends to disengage the viewer from the kind of visualization of a machine's performance that is needed to assess a plan. An impression is created of bodies suspended in space and located outside of time. Information about the dynamic interaction of a machine's components during its performance requires visual exploration of exposed and hidden surfaces, anticipating what would happen under various experimental conditions. For such an exploration, a literate observer must retrieve multiple views of surfaces from a variety of perspectives. Of course, drawings in orthographic projection provide valuable information about geometrical and physical properties. However, if designers want to show clients how the machine operates, alternative projection systems are needed. A viewer actively attends to the subtle movements of components, as if one is placed in a scene and becomes attuned to possible changes in a visual landscape. Assessment of a design plan requires apprehension of some surfaces that would come into sight, and other surfaces that would pass out of sight, based upon a viewer's imaginary movement.

Obliquity typically distorts some surfaces of objects in ways that portray other surfaces. The rule for such a technique is to draw the front face as a true shape but foreshorten the side or top faces. A small change of position for the object or viewer does not change the number of faces, edges, or corners that the viewer can see (Dubery and Willats 1983, p. 44). Many rectangular objects in pictures are easily recognized when they are shown in a foreshortened position. Rodin gave movement in his busts by giving them a certain slant, or obliquity, suggesting an expressive direction (Arnheim 1964, p. 344).

Oblique projection is evident in the work of Renaissance designers. Consider Leonardo's drawing of a machine used for boring solid bodies, seen in Figure 10 (Baroni 1956, p. 503).

In looking at this figure, a viewer imagines being positioned to the right and slightly above the machine's wider side. A sense of aesthetic movement is generated as a viewer imagines entering a scene from the top right of the

FIGURE 10. Leonardo's drawing of a machine relies on aesthetic movement of figures

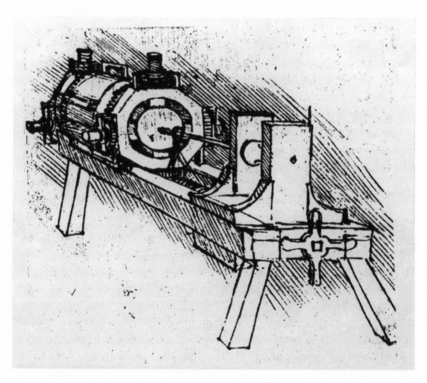

wider side, looking across the face of the machine's wider side, and exiting the bottom left. Leonardo's sketch of the boring machine exhibits both horizontal obliquity and vertical obliquity. In a horizontal oblique projection, the oblique angle runs along a horizontal axis, while the vertical axis remains perpendicular to the plane. Again, obliquity in line drawings creates a sense of possible change in the angle of projection from the environment to a picture plane. We can imagine a change of perspective caused by an imaginary change of position to the left or to the right of the figures. From such a change, surfaces that appeared foreshortened from one perspective will now be exposed in their true shapes, and simultaneously other surfaces in true shape become distorted. This is commonplace in folk art, Byzantine art, and children's drawings of houses (Willats 1997, p. 48). In vertical oblique projection, the oblique angle is established between the projection rays and the vertical axis of the picture plane, not the horizontal axis, as if one enters the bottom left and exits the top right (Dubery and Willats 1983, p. 33).

Notice the similarities between Leonardo's two drawings in Figures 8 and 10. He masterfully enhances the dynamic features of the scene by generating horizontal obliquity and vertical obliquity. A viewer imagines standing above and slightly to the right of the action. Figure 10 exhibits a sense of aesthetic movement by overlapping the faces of the machine's components, engaging a viewer in the following perceptual problem: How can a viewer distinguish between the actual faces of a machine's components and the possible faces that would appear from a different point of observation? Some faces of components are exposed while others are hidden from a viewer's perspective. An impression of depth is enhanced by the contrast between positive space of the surface of the forehead and negative space of an "empty" background.

In contemporary engineering, an oblique projection shows the front face of the material body in its "true" height and width, while two faces of the material body are projected at an oblique angle, between 20 degrees and 60 degrees. Designers often supplement their plans with illustrations that traverse the picture plane obliquely. For example, in a schematic illustration of absorption spectrometers, as shown in Figure 1, oblique projection enhances information about the functional relationships among components of the instrument. In Figure 11 (p. 56), a schematic illustration of a two-filter wheel is shown in horizontal oblique projection, as if a viewer is placed slightly to the right of the wider side of the wheel (Coates 1997, p. 447).

In this figure, horizontal obliquity creates a sense of movement in a design space from the immediate perceptions of the actual to an apprehension of

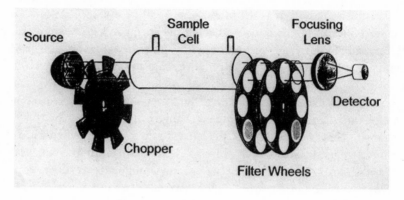

FIGURE 11. A schematic of a two-filter wheel, multi-wavelength filter photometer

possibilities. A skilled reader of line drawings transcends the limits of momentary perceptions by apprehending possible scenes.

In sum, the ancient notion that thinking is a form of vision finds a contemporary expression in the designs of analytical instruments. In their graphic designs of machines, engineers frequently resort to a language of lines, shapes, and pictorial images. Schematic illustrations are commonly used for showing how a device should perform under ideal, or typical, conditions. Assuming visual literacy, icons of design are coded as a hieroglyphics of pictorial symbols. The icons convey information about the unknown through a process similar to our experiences with occluding edges. An aesthetic tension emerges as a pivotal component of iconic modeling: a sense of movement of images, shapes, and forms underpins iconic modeling of design. Contrary to the algorithmic operations of rational problem solving, modeling rewards designers who can work with multiple images from different visual perspectives. A successful designer creates in viewers an experience of visual tension from a juxtaposition of symmetries and asymmetries. The asymmetries in design are revealed in images of such patterns.

In chapter 5, I turn to a historical study of the natural philosophers of the seventeenth century, exploring technological innovations of laboratory devices that produced a scientific revolution. Natural philosophers engaged in laboratory research by immersing themselves in the mechanic's trade, learning from artisans how to manipulate certain materials for the production of optical devices, and finding professional advancement by working with metals, wood, and glass in the shops of instrument makers. Robert Hooke in particular found philosophical reward in his design and production of compound microscopes. Through such activities, he affirmed the principles of a mechanistic universe, whose properties are evident not only in artifacts of an instrument manufacturer but also in the smallest particles of material bodies.

5

MICROSCOPES, MACHINES,

AND MATTER

True [experimental] philosophy begins with Hands and Eyes, proceeds
through Memory, [and is] continued by Reason, coming back to Hands
and Eyes again.
—Robert Hooke, *Micrographia*

Sir Isaac Newton, Robert Boyle, Robert Hooke, and other natural philosophers of the seventeenth century found a universe without a specific purpose,
a world that was neither benevolent nor cruel, neither beautiful nor ugly.
Created by the Grand Watchmaker, the universe in their view is driven by
Cosmic Machines that are hard, cold, colorless, silent, and dead—completely
unaffected by man's actions, achievements, and ambitions. The gloriously
romantic world of Dante and John Milton, which set no limits on the imagination of man as it played over space and time, was swept away (Burtt 1992,
pp. 238–39). Familiar qualities of experience consisting of a world rich with
color and sound and filled with purpose, love, and beauty were banished to
minute corners of the brain. Such qualities were either minor effects of the
Cosmic Machine of nature or were summarily dismissed as unworthy of serious scientific detection. Man was reduced to a puny, irrelevant spectator.

Natural philosophers bestowed honor upon those who could disclose
nature's secrets. The discovery of powerful new instruments, such as the
compound microscope, telescope, air pump, and barometer, promised to
reveal the inner workings of the Cosmic Machine. Man was no longer the
axis about which the cosmos revolved, but now, for the first time, he was able
to look inside the divine clockworks. The medieval notions of up and down
in a spiritual cosmology found no place in experimental research. Although
experimenters looked up when using telescopes and down when using microscopes, they found no preferred center of the universe. The Moderns
sought to expunge from experimental research the biases that might arise
from cognition. The new instruments were designed to replicate and enhance
God-given sensory capacities, with human vision serving as a prototype for

telescopes and microscopes. We read in the philosophical writings of the day that such devices augmented human sensory capacities and removed some potential biasing aspects. The Moderns recognized their new tools properly as philosophical instruments, providing experimenters with access to truths of the material world.

The Moderns had resurrected a form of anthropocentric orientation to experiments that relied upon human organs as a basis for instrumental design. But we must guard against overstating the historical case for such reliance. Although designers of optical devices depended on rigorous studies of human vision, originating in the work of the ancients, most natural philosophers of the modern era explained sensory perception through mechanical interactions among atoms. The instrumental techniques mimic a machinist's skills rather than an observer's sensory capacities. The dominant prototype for instrument makers was not the observer's eye but the machinist's eye and hand. The eye itself was one of nature's machines.

In this chapter, I examine the compound microscopes of the modern era, identified by designers as philosophical instruments. Hooke's fame in seventeenth-century natural philosophy arose from spectacular achievements through the use of microscopes that he developed. The rationale for such a use presupposes a particular notion of a scientific object. A scientific object must be amenable to a mechanic's sincere hand and faithful eye for slicing, dissecting, or simply inspecting a body under a microscope. A study of machinists' skills offers insight into the causal mechanisms that are exploited in the microscope's design and detected in its use.

WHAT ARE PHILOSOPHICAL INSTRUMENTS?

Two categories of instruments were adopted before the seventeenth century: mathematical and philosophical. Since the time of the ancient geometricians, mathematical devices aided researchers in numerous practical affairs, such as surveying, navigation, and military fortifications (Warner 1994, pp. 67–68). The sector, for example, was a hinged ruler used freely in gunnery and surveying (Price 1957, pp. 627–28). Mathematical instruments were constructed from wood and brass, with ivory and leather used for decoration. Such devices were used to weigh, measure, or otherwise attach numbers to properties of nature. Special skills of engraving became pivotal: craftsmen had to cut lines and scales on a flat metal plate, usually by hammering and filing. Because sixteenth-century manufacturers lacked sophisticated methods for determining graduating scales, the accuracy of such scales was often sacrificed in production (pp. 622–24). But such devices were still used for the businesses of traders, merchants, seamen, carpenters, and surveyors.

By contrast, the reward for using instruments of natural philosophy was not found in the practical affairs of commerce and the military but rather in the discovery of true laws of nature. The use of these instruments implied authority and prestige, often boosting a scientist's status up the social or intellectual ladder (Warner 1990, p. 84). Natural philosophers designed their new devices to reveal a world of immense physical power, disclosing otherwise hidden processes that were causally responsible for observable events. For such disclosures, researchers employed compound microscopes to interact with nature at a higher degree of resolution. The early microscopes of the 1650s were designed to agitate a body's minute particles. The resulting optical processes were responsible for visual images observed through the eyepiece. Mathematical instruments, in contrast, were passive devices, devoid of any mechanism to transform material bodies and incapable of revealing underlying causal processes responsible for natural events (Warner 1994, p. 68).

In their search for truth, natural philosophers were increasingly drawn to the shops of instrument makers. The best natural philosophy could be learned from machinists who produced telescopes and microscopes. The production of telescopes and microscopes in the 1650s required lens-grinding skills that were familiar to the spectacle-making industry (Price 1957, pp. 632–33). Such skills included metalworking, woodworking, glassworking, and tube-making. In preparation for his optical experiments using prisms, Newton devised new techniques for producing lenses "free from Bubbles and Veins" with "truly plane" sides (Schaffer 1989, pp. 93–94). As an expert lens-worker, Christian Huygens expended considerable time and energy developing sea-clocks, which required both theoretical insight and practical skills that he hoped would change the art of navigation (Bennett 1986, p. 7). More than ever before, natural philosophers identified their professional status through the skills associated with manipulating such devices, and wondrous discoveries were brought about by the use of such instruments. The enthusiasm for these devices prompted a large demand, which was satisfied through mass production (Price 1957, p. 628). At the beginning of the sixteenth century, Nuremberg and Augsburg were well-known centers for the craft of instrument-making. By the end of the sixteenth century, various centers for instrument-making had opened in England, France, Italy, Germany, and the Low Countries (pp. 620–22).

Of course, progress in microscopy relied on advances in spectacle-making. Roger Bacon, among many others in the twelfth century, explored techniques for using convex lenses to improve vision. Designed to compensate for visual impairments, the early spectacles had convex lenses, ground to curves of a small radius. By the time of Bacon's death in 1294, spectacles were produced in considerable quantity by a burgeoning glass industry in Venice (Charleston and Angus-Butterworth 1957, pp. 229–31).

By the mid-1600s, instrument makers discovered the visual power that resulted from combining lenses in a single microscope. In one of the monumental advances of modern science, the optics of the compound microscope was revealed. Although the issue of historical originality is not settled, many scholars credit this discovery to a Dutch spectacle maker named Jans Janssen who, along with his son Zacharias, stumbled upon the technique quite by accident, possibly from the chance act of holding two lenses together (Bradbury 1967, pp. 21–22). Before 1650, telescopes were made of wood, brass, lenses, and occasionally ivory and leather for decoration. The first use of lenses for telescopic vision was probably developed by the Janssens' work in 1608 and was immediately put to military use (Charleston and Angus-Butterworth 1957, pp. 230–31). In its function during military campaigns, the telescopic lens began as a device that is not merely "philosophical." Learning of this Dutch invention, Galileo converted this device to study planets and stars, leading to great excitement with the discovery of Saturn's rings, Jupiter's satellites, spots on the sun, and mountains on the moon. Galileo is known to have converted the telescope, with its concave lens, into a microscope (H. Brown 1985). Marcello Marpighi, one of the most original microscopists of the era, performed extensive examinations of the human lungs and the circulation of blood.

René Descartes was fascinated with optical technologies and was drawn to the shops of the glass industry in Holland, his adopted homeland. In the Ninth Discourse of *Optics,* he provides illustrations of an immense instrument, recognized as the earliest known drawing of a compound microscope (Descartes 2001, p. 157).

Because the angle of the device in relation to the heavens suggests work in astronomy, this illustration, shown in Figure 12, appears to show a telescope. But the placement of the specimen clearly indicates an instrument of microscopy. The specimen is illuminated by light, probably from the sun, which reflects off of a hyperbolic mirror. The microscope includes a convex objective lens and a single concave lens acting as an eyepiece (Bradbury 1967, p.

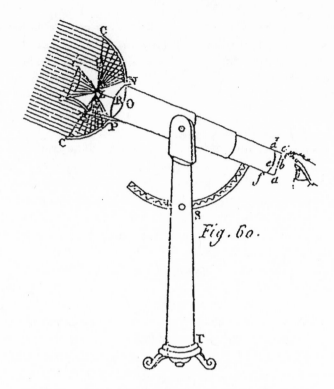

FIGURE 12. Descartes's sketch of a compound microscope

19). (The images presumably provide evidence of a body's primary qualities, understood by the geometrical properties of shape, size, and quantity.)

Explorations by natural philosophers in the mechanical arts of instrument-making cannot be explained by dividing their efforts into two clearly distinct domains, one philosophical and the other practical/mathematical. Machinists contributed much more to natural philosophy than constructing new apparatus. Hooke championed the cause of blending experimental research with the mechanical arts. The standards for inquiry rest on the techniques for manipulating material bodies, as found in the machinists' craft. By the beginning of the eighteenth century, the sharp distinction between the two categories of instruments began to erode. One reason was that many "philosophical" instruments were equipped with so-called mathematical components, such as graduated circles. Conversely, some "mathematical devices" were given telescopic sights. By 1800, some Englishmen identified themselves as "mathematical and optical instrument maker" or "mathematical, optical, and philosophical instrument maker" (Warner 1990, pp. 84–85).

PHILOSOPHERS AS MACHINISTS

One driving motivation for inventing philosophical instruments was the need to overcome the damage from unbridled speculations of metaphysicians from an earlier era, especially the Medieval Schoolmen (Hooke 1961, preface). Reliance on images produced by experimental techniques suppressed the contaminating effects of metaphysical judgments. These devices were designed to increase the exactness of observation or to extend vision to otherwise inaccessible regions. Telescopes magnify the eye's natural powers, the ear trumpet improves hearing, and tactile senses are simulated in the barometer and thermometer.

The philosophical apparatus of the seventeenth century did not receive immediate approval. Critics charged that the images produced by microscopes, for example, were mere artifacts of the apparatus and did not reveal nature's true properties. Why divert resources to some useless collection of metals, wood, and mirrors? Huygens lamented that microscopes produced mere illusions. In the mid-1600s, the famous experimentalist Henry Baker went so far as to challenge the honesty of the experimenters who used optical devices. After observing some of the first experiments using microscopes presented to the Royal Society, Baker questioned whether the experimenter lied about his observations or declined to use his best instruments (C. Wilson 1995, p. 225). Even Boyle, whose development of the air pump greatly advanced instrumental technique, charged that the visual images of the microscope at best provide surface effects without revealing the underlying "alphabet" of nature. In empirical studies of color, greater magnification and a sharper image do not expose deeper truths, because more images amount to added mystification (C. Wilson 1995, p. 228).

Thomas Hobbes questioned the wisdom of using any optical instrument in empirical research and predicted that all the fuss about the new technologies would soon pass, like a frivolous parlor game. He claimed that such devices actually confound researchers by producing mere artifacts of technique instead of accurate information about real-world properties. Serious-minded researchers should not be deceived by the stunning and sometimes beautiful images of optical devices. Declaring that the contaminations produced by such devices would be exposed when researchers compare instrumental findings to observations made with unaided senses,

Hobbes implored philosophers not to be swayed by the mechanical trick-sters who produced spectacles for amusement rather than secrets of nature (Shapin and Schaffer 1985, pp. 128–29).

Even in the eighteenth century, after acknowledging the stunning results from optical instruments, John Locke vigorously objected that the use of optical devices in research is ill-founded, reflecting a confusion between words and things. When microscopes are used for an experiment, Locke said, the visual images are confused with statements. Such devices are useless to experimenters.

Of course, many philosophers found great promise in the new devices. Descartes, George Berkeley, and Immanuel Kant recognized the necessity of using such devices in research. The primary mission of natural philoso-phers centered on hopes for a more profound understanding of nature. All empirical events, including those that produce images on the eyepiece of a microscope, were thought to be surface effects of the configuration and motion of basic units, such as corpuscles or their aggregates (Capek 1961, p. 79). For microscopists, the images from microscopes were causally gen-erated by a continuous sequence of events, from the agitation of a body's minute particles to the contact of luminous bodies on the lens of an eye. Each philosophical instrument was defined by its capacity to receive mo-tion and to communicate motion to other bodies, according to principles of mechanics.

In this rationale for philosophical instruments, some commentators find evidence of an anthropomorphic orientation to science (Kutschmann 1986). Seventeenth-century scientists constructed substitutes for almost all human senses. Not surprisingly, empirical studies in the anatomy of the naked eye were pivotal to the design of artificial surrogates, such as the lens, magni-fying glass, spectacles, telescope, and microscope (p. 107). In Renaissance astronomy, the appeal to bodily analogies was quite common, especially to identify the beauty and symmetry of the cosmos (Kemp 1996, p. 60). Of course, the problem of relativity of observation generated a paradox: How can observations be trustworthy when appearances of shapes and motions depend upon an observer's position? Subjectivity of perception was avoided by the principles of vision and the geometry of perspective, which contrib-uted to man's role as an observer of the heavens (p. 64).

Although microscopes were designed to enhance sensory capacities, the rationale for using such instruments given by designers centered on the me-chanical interactions with minute particles, analogous to a machinist's ma-nipulations of bodies. The naked sensory organs themselves were understood as God-given machines. An organ is a prototype for an instrument, and the physiology of the eye was the model for the operation of artificial devices designed to enhance vision. The mechanic's skills in handling, changing, and producing material entities function as exemplars to explain an instrument's powers of sensory magnification. The experimenter transforms certain properties of a material body rather than passively observes its empirical attributes. This theme can be revealed by contrasting philosophical apparatus with another category of instruments: the instruments of natural magic.

Most philosophical instruments of the modern era were developed in earlier forms for purposes of natural magic. The earliest known sketch of a telescope is by the natural magician Giambattista Della Porta. Boyle learned of the air pump from reading the work of the natural magician Gaspar Schott. Newton got his prisms at a fair where they were sold as instruments of natural magic. Natural magicians produced startling effects through trickery from hidden mirrors. The goal was to emulate the wonders of nature for entertainment or practical benefits without exposing the underlying causes of such events (Hankins and Silverman 1995, pp. 4–5). Magicians sought to beguile observers with bewildering experiences that disguised their true causes.

In one familiar technique, magicians would substitute one sense for another. Kinesthetism is the ability to conceive an object interchangeably or concurrently through visual, verbal, mathematical, or kinesthetic techniques (Root-Bernstein 1991, p. 335). For example, Louis-Bertrand Castel designed the ocular harpsichord to display colors that corresponded to certain tones. Castel tried to paint sounds by corresponding color with pitch, based on an analogy between the group of seven spectral colors and the seven tones of a musical scale. Again, the devices of magic were truth-concealing, intended to disguise the true causes of a spectacular event.

Do philosophical instruments disclose nature's secrets or are they sources of deception? Designers promised much more than enhancements of God-given perceptual skills. The rationale for using such devices offers insight into experimental inquiry and requires an understanding of the causal properties that are responsible for observable events. How could natural philosophers persuade critics that the visual images produced by optical instruments are products of nature's truth rather than artifacts of the new technology? As energetic participants in the design of microscopes, many natural philosophers found their work increasingly drawn to the mechanical arts, representing a profound challenge to the distinction between high science (theoretical explanation) and low science (mechanical arts).

In medieval societies, the artisan was not typically engaged in natural philosophy and not expected to disclose truths about the causal properties of the material world. The clock maker worked with wheels and springs and was not concerned with theories of corpuscular interaction. But this medieval distinction between the scholar's work and the craftsman's skills seems sterile with the seventeenth-century construction and use of philosophical instruments.[1] As James Bennett documents, this appeal to the craft of instrument making put into practice a mechanics' philosophy (1986). Since a material body was reducible to machines, the methods for understanding nature should be drawn from the practical techniques commonly associated with the mechanical arts. Man-made machines conform to the same principles as the machines of nature. The cosmic metaphysic was quite relevant here: the artifacts produced by hammers and forces conformed to the same universal principle of materiality as any work of God (Baigrie 1996, p. 102). We find in the craft of experimental machinery certain skills that contributed directly to an experimental philosophy. The traditional separation between high science (from the theories of philosophers) and low science (from the products of artisans) withered away with the development of the new devices (Bennett 1986, p. 6).

The mechanics' philosophy is probably best epitomized in the experimental insights of Hooke, known as one of the premier English microscopists of the seventeenth century.

ROBERT HOOKE'S MICROSCOPY

Hooke was the first natural philosopher to systematically record experimental discoveries using compound microscopes, stimulating tremendous enthusiasm among natural philosophers of Europe. In his early days of research, Hooke was influenced by John Wilkins, a member of the Gresham College. Known as the center for practical mathematical professionalism in England, Gresham housed informal groups of scholars and eventually gave birth to the Royal Society (Bennett 1986, pp. 22–23).

Hooke showed considerable talent in the craft of lens-grinding, prompting his design of instruments for experimental inquiry. His diary records almost daily visits to manufacturers for the purpose of refining lenses (Price 1957, p. 630). He discovered a new kind of glass, which he used to develop new telescopes and microscopes. With their greater magnifying power, these new telescopes could now disclose a visible world of heavens and stars, possibly including the discovery of living creatures on the moon or other planets. But Hooke's true masterpiece was achieved in microscopy.

In one of the great works of science literature, *Micrographia or Some Physiological Descriptions of Minute Bodies Made by Magnifying Glasses with Observations and Inquiries Thereupon* (1961), Hooke provides detailed illustrations of microscopic discoveries. Exhibiting his training in fine arts, he deftly illustrates the edge of a razor, point of a needle, and moss on leather. In fact, *Micrographia* functions as a manual for using the compound microscopes that he designed (see Figure 13).

Insects, such as lice, gnats, and flies, were placed under the microscope, and *Micrographia* includes a wonderfully detailed sixteen-inch diagram of an insect (1961, p. 210; see Figure 14).

The surface of a cork reveals small pores, or cells, as Hooke called them. This may be the first known use of the term "cell" in biology (Hooke 1961, p. 55).[2] After learning of Hooke's discoveries, the Royal Society solicited him in 1663 to display at least one microscopic observation at every meeting (Bradbury 1967, p. 39). His contribution to the Society is legendary.

FIGURE 13. Illustration of a compound microscope by Robert Hooke

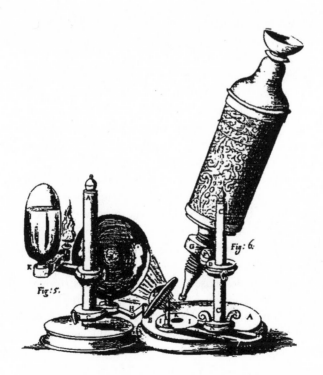

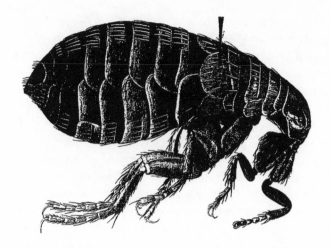

FIGURE 14. Hooke's drawing of insect as observed through a
compound microscope

Hooke admitted to two serious defects in his microscopes, adding to others' suspicions about the veracity of the visual images. First, a slight blurring of the image occurs because of bending or refraction of light rays. He failed to discover a satisfactory means for overcoming this defect (Bradbury 1967, p. 42). Second, chromatic aberrations occur when the image includes colored edges. Hooke realized that single-lens microscopes were less affected by these obstacles than were compound microscopes, although both kinds were used in his experimental studies (Hooke 1961, preface).

Although the new optical instruments functioned as surrogate sensory organs, the rationale for their design was explicitly mechanistic. The true source of his experimental philosophy was not modeled after an observer's eyes but rather a mechanic's hands and eyes. Just as machinists used handheld tools to craft their raw materials, experimenters used optical instruments to augment their ability to physically manipulate material bodies. Hooke searched for the universal properties of nature's machines in the mechanical arts in an effort to persuade potential critics that the images produced by microscopes are the observable effects of motions and interactions of small, invisible particles. Based on the ways that instrument makers mimicked mechanistic processes of nature, he anticipated the development of instruments to improve hearing, tasting, smelling, and touching (Hooke 1961, preface). For example, the normal range of hearing could be enhanced by a device called a stethoscope that could detect minute sounds of animals and vegetative bodies (Gouk 1980, pp. 579–80).

In their widely heralded book *Leviathan and the Air-Pump: Hobbes, Boyle, and the Experimental Life,* S. Shapin and S. Schaffer characterize Hooke's rationale for the compound microscope through empiricist standards for knowledge (1985). We read that for Hooke, the power of new scientific instruments rests on their capacity to enlarge the dominion of the senses, revealing spectacular scenes to those skilled in instrumental techniques. Hooke implores experimenters to use such devices to overcome the "infirmities" of the senses and to enlarge the range of empirical attributes accessible to experimenters (Shapin and Schaffer 1985, pp. 36–37). Of course, the metaphysical excesses of the Medieval Schoolmen must be avoided.

This reading of Hooke's experimentalism seriously distorts his rationale for microscopic research. Conspicuously absent from the interpretation by Shapin and Schaffer is a reference to Hooke's mechanistic explanation of instrumental techniques. Four themes can be gleaned from Hooke's microscopy. First, one of the widely held fictions about modern science is that the new optical instruments of the seventeenth century were passive devices that provided quasi-transparent windows to empirical attributes. In this view, unobservable attributes remain inaccessible to experimenters even with the use of instruments. But Hooke explicitly implored researchers to use compound microscopes to reveal the workings of otherwise undetectable (unobservable) machines of nature. The natural world is a series of smaller machines embedded in larger ones. Motions of machines at one level are produced by machines at a deeper level, comprising the wheels, engines, and springs of the microworld. Each material body is endowed with powers that are causally responsible for detectable effects. Every tool is an agent for change, that is, a source of movement that is responsible for certain occurrences. The success of artificial devices is explained through the underlying mechanisms that are causally responsible for producing such events.[3] The ability to discover any natural machine requires work by designers and researchers. The power of a compound microscope to reveal truths rests on the designer's exploitation of mechanisms that are causally responsible for the images at the eyepiece. For example, what are the causal agents responsible for the fact that some fluids readily unite with, or dissolve within, others? How can we explain fermentation by yeast? How can the infection of one man lead to the destruction of thousands of others? Discovery of nature's causal mechanisms must be included in the mission of experimenters, even though such mechanisms are "secretly and far removed from detection by the senses" (Hooke 1961, p. 47).

Second, conjectures about any natural body can be drawn from analogical associations to the workings of artificial machines. Without any direct, or even instrument-mediated, observation of the deepest regions of the universe, experimenters must resort to hypothesis and reasoned speculation. Natural philosophers must look to machine shops for insight into the workings of the universe. Although mechanics is a practical affair, partly physical and partly mathematical, Hooke drew philosophical inspiration about the truths of nature from the discoveries of machinists. Even human diseases can be diagnosed by discovering "what Instrument or Engine is out of order" (Hooke 1961, p. 39). Natural philosophers should become acquainted with the machinists' techniques of hammering, pressing, pounding, grinding, cutting, sawing, filing, soaking, dissolving, heating, burning, freezing, and melting (p. 60). Of course, these techniques include a skilled manipulation of material bodies and typically require physically connecting bodies through construction of joints and hooks. A watch offered special insight into the workings of nature, with the balance beating, wheels running, and hammers striking. By analogy, "[It] may be possible to discover the motions of the Internal Parts of bodies, whether Animal, Vegetable, or Mineral, by the sound they make" (p. 39).

Third, for Hooke any machine of the natural world is characterized by its agent capacities to produce movement and its reagent properties to be moved by other bodies. All matter is reducible to the physical interactions of tiny machines. Any machine is knowable through its physical powers to manipulate, agitate, or transform another machine and can be altered itself. Explained through the movement of its component parts, a fluid is "nothing else but a very brisk and vehement agitation of the part of a body . . . [and] the parts of a body are thereby made so loose from one another that they easily move any way, and become fluid" (Hooke 1961, p. 12). The power of any machine, including laboratory instruments, is explained by the workings of a causal mechanism, comprising a system of entities with capacities to generate movement in, or to be moved by, other bodies. We can say that the conditions of agitation are causal events and the resulting occurrences are effects.

Fourth, to present their findings about workings of a machine, whether artificial or natural, experimenters must resort to visualization techniques. For a proper experimental philosophy, "the intellect should, like a skilful Architect, understand what it designs to do" (Hooke qtd. in Waller 1705, p. 18). Designers often resort to the artist's skills for the purpose of registering the history of a particular investigation. Through schematic illustrations, designers offer a vividness of presentation that invites vicarious participation, encouraging viewers to distinguish the fleeting events of the laboratory from the experiment's "elements." The complex workings of the smallest wheels, engines, and springs can be visually depicted.

Hooke implored experimenters to develop drafting skills, not only to express ideas, but also to better inform and instruct others. "Many things cannot be made as plain by a large description in words as by the delineation of them in a quarter of a Sheet of paper" (Hooke 1961, p. 20). Verbal discourse was necessary but by itself too confining for showing experimenters the workings of machines. We need visual narratives to show the workings of machines, both natural and artificial. Of course, the smallest machines of the world are not what they seem to be from their attributes at the surface. The lattices in an insect's eye were conveyed visually and not just described verbally. Hooke skillfully integrated text and graphics to show how his new instruments worked. In his visual narratives of experimental procedures, we see how he glued a fly to a feather to observe its attempts to fly, how he froze a fly and then gently thawed it by the fire, and how he got a fly drunk on wine and then sobered it up (Harwood 1989, p. 144). The plates in his *Micrographia* are detailed works of art, representing a composite of many observations.

HOOKE'S EXPERIMENTALISM

Hooke faced the following daunting task: how to demonstrate that his instruments interacted with a body's minute particles in ways that produced accurate images visible to the naked eye. His rationale for the compound microscope includes an account of three kinds of movement at the microscopic level: (1) movement of a material body's minute particles, (2) action of luminous bodies producing optical effects, and (3) movement of the retina of the human eye. These movements show the workings of nature's machines.

1. Every gas, fluid, and solid is understood mechanistically through the interaction of the minute particles that "swim and move" against one another in a continuing ether. Each particle is itself a small machine. The material essence of any machine is found in its power to displace—and to be displaced by—other bodies. That is, each machine has a capacity to receive motion from bodies and to communicate motion to bodies. Of course, a particular volume of air requires fewer vibrating particles than does the same volume of a solid. Properties of a fluid are explained by the agitation of particles due to the pulse of heat (Hooke 1961, p. 12). A solid body is least subject to

deformation because of the short vibrations of its minute particles. As the hardest body known in the world, a diamond is comprised physically of particles whose vibrations are shorter than any other body. Consequently, a diamond is least subject to deformation from contact with other bodies (pp. 55–56). The entire physical universe consists of the interaction of material bodies, each of which is characterized by the power to be moved and to displace other bodies.

Hooke's studies of light, heat, and sound show Descartes's influence. Hooke endorsed Descartes's idea that the motion of particles is entirely local, characterized as a tendency of a body to change its position in relation to other bodies. But Hooke discovered that the ethereal medium between material bodies is not continually whirling in a series of vortices, as Descartes believed. For Hooke, the material medium is static and remains still while material bodies circulate through it (Gouk 1980, p. 585).

2. Hooke believed that visual images are produced when a material body emits radiation that travels through a medium to the organ of sight. By the seventeenth century, the old (medieval) rivalry between extramission theories of optics and intramission theories had been settled. Euclid advanced an extramission theory according to which vision occurs when radiation is emitted from the eye and sent to "feel" the properties of visible objects (Lindberg 1992, p. 341). An early advocate of intramissionist theory was the thirteenth-century Islamic natural philosopher Ibn al-Haytham. From his influence, intramission theory achieved a new level of success for explaining the physical contact between object and observer through intramitted rays. The magnification of glass spheres is explainable through the following causal process: radiation is emitted from the visual body and transmitted through a medium, then impinges on the convex surface of the eye. Al-Haytham created a new tradition that integrated optics with the anatomical and physiological claims of medicine. The intramissionist character of vision would never be seriously doubted again (p. 348).

Hooke offered mechanistic explanations of the ways in which sensory images are acquired by the eye with or without the aid of optical devices. Such images are effects of the pulsating action of a body's minute particles. This action is then communicated through the progression of rays, moves in a medium, and enters the lens of an eye (or a microscope). For example, a body's color can be explained by "working backward" from the images to the operations of the naked eye, to the optics of the luminous medium, and eventually to a material source of illumination (Hooke 1961, p. 7).

3. Hooke endorsed the view that the eye itself is a natural machine en-

dowed with powers to convert radiation into visual impressions. Of course, he was not the first philosopher to do so. For Galileo, the images emanating from the naked eye could be explained by the same mechanical laws that apply to any optical instrument. But optical devices of natural philosophy are not susceptible to the troublesome defects of "natural" vision. The eye, in contrast to a philosophical instrument, produces certain distorting rays that must be evaluated and corrected (H. Brown 1985, pp. 499–501). Descartes also endorsed the view that the living eye was an optical instrument, but Hooke rejected Descartes's explanation of optical phenomena largely because Cartesian optics failed to explain the ability to distinguish colors (Hooke 1961, pp. 60–61).

According to Hooke, two light rays fall on the cornea and then converge into two points at the bottom of the eye, as indicated in Figure 15 (Hooke 1961, p. 60).

The rays enter the eye obliquely, causing them to be refracted in the cornea. As a result of this oblique progression of the rays, one side of the pulse

FIGURE 15. The optics of color according to Robert Hooke

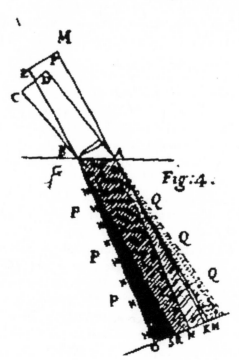

arrives at the lens and penetrates into the refracting medium before another side of the pulse. This difference in the propagation of rays explained the appearance of colors. Depending on which rays touch the retina first, one cornea will receive an impression before another. Blue is an impression on the retina caused when the "weakest part" of the ray precedes the strongest part, whereas red is an impression whose strongest part precedes its weakest part (p. 64). Unfortunately, Hooke does not provide a clear conception of pulse strength (Sabra 1967, p. 255), and so his theory proves to be unsatisfactory.

ANALOGIES TO SOUND

We sometimes understand instruments as created in our own image. Beginning in prehistoric times, ideas about the human body have inspired the invention of tools. The compound microscopes of the seventeenth century were designed by extending, enhancing, or exploiting capabilities with which we are endowed and intimately familiar. While Hooke designed his compound microscopes as surrogates to a machinist's hands and eyes, he also understood human organs as God-given machines. Hooke's analogies of visual images to sounds were intended to reveal the universality of the idealized machine. The causal mechanism underlying musical tones was also responsible for the reliability of images from compound microscopes. Such analogies were developed to garner professional support for the new philosophical instruments.

The power that compound microscopes confer on experimenters was no mystery to Hooke. When using a compound microscope, an experimenter handles something perceptible so as to manipulate something imperceptible. As shown above, the compound microscope achieved reliable results because of the workings of nature's machines for producing optical rays. Hooke understood the mechanisms underpinning the compound microscope through analogies to the principles of acoustics. For Hooke, the motion of rays is caused by an internal periodic vibration of minute bodies, analogous to action of a musical string. His analogical associations of light to sound were intended to explain the power of instruments. In particular, a light ray is like a string stretched between a luminous body and an observer's eye. The mechanics of strings are universal. The three properties that determine a musical tone are the string's length, tension, and thickness. Found in vibrating particles, these properties are characteristic of the productive capacities of any minute body, so that the universal principles for generating any sensible event are revealed in acoustics. The explanation of the devices of wood, glass, and metals is found in the universality of all machines. Hooke writes: "To which three properties in strings [length, tension, and thickness] will correspond three properties also in sand, or the particles of bodies, their matter or substance, their figure or shape, and the body or bulk. And from the varieties of these three, may arise infinite varieties in fluid bodies, though all agitated by the same pulse or vibrative motion. And there may be as many ways of making Harmonies and Discords with these [fluid bodies], as there may be with *musical strings*" (1961, pp. 15–16; emphasis in original). The vibrative motion of musical strings that generates possible tones is a prototype for the vibrations of all material bodies (Gouk 1980, p. 580). Hooke found empirical confirmation for this construct during his work as Boyle's laboratory assistant. For example, a vessel of water will undergo a trembling motion when influenced by certain tones in its vicinity. In like manner, sound can also cause motion of flour in a container.

In his most original theoretical work, Hooke proposes a principle of congruity and incongruity, based on analogies to music (Gouk 1980, p. 585). According to this principle, every gas, fluid, and solid is endowed with a capacity to unite with certain other bodies, depending on their physical attributes. "[Particles] that are all *similar*, will, like so many *equal musical strings equally stretcht, vibrate* together in a kind of *Harmony or unison;* whereas others that are *dissimilar* . . . will, like so many *strings out of tune* to those unisons . . . make quite *differing* kinds of *vibrations* and *repercussions* [as to] *cross* and *jar* against each other" (Hooke 1961, p. 15; emphasis in original). Congruity refers to a body's tendency to mix "harmoniously" with another part of the same body or with some other similar body. Particles of "like bigness, and figure, and matter" will adhere or "dance together." In this way, bulky bodies are kept together by congruity of their parts. Heterogeneous bodies, on the other hand, are averse to such adherence (p. 15). Because of their incongruity, certain bodies are incapable of mixing, as when bubbles of air remain unmixed in relation to the surrounding water.

Hooke's analogy of light to sound found support from an unexpected source. One controversy that bitterly divided Hooke and Newton centered on the explanation of colors. For Hooke, Newton's atomistic explanation of the color spectrum was flawed without a commitment to an ethereal medium through which corpuscles move. But eventually Newton adopted Hooke's analogies between optics and acoustics. Just as a tone requires the vibrations of surrounding air, color is determined by the vibrations of light rays through an ethereal medium. In Query 13 of his *Opticks,* Newton writes: "Do not several sorts of Rays make Vibrations of severeal bignesses, which according to the bignesses excite sensations of severeal Colours, much after the manner that

the Vibrations of the Air, according to their severeal bignesses excite Sensations of severeal sounds? And particular do not the most refrangible Rays
excite the shortest Vibrations for making a Sensation of deep violet, [and]
the least refrangible the largest for making a Sensation of deep red?" (1952,
pp. 345–46). So, red is produced from the strongest vibration and deep violet
from the weakest, illustrating how "the Analogy of Nature is to be observed"
(Newton 1959, p. 376). In a letter to Henry Oldenberg, Newton analogizes the
color spectrum to tones on the D-string of a violin (see Figure 16).

The "spaces" among the seven colors are analogous to the seven tones of

FIGURE 16. Newton's analogy between color spectrum and
scale of notes

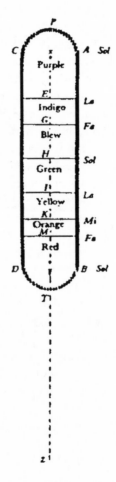

a rising major scale, which in turn are explained by the mechanistic properties of the vibrating ether (Newton 1959, p. 377 and fn. 14).

Again, for Hooke the power of any machine, whether natural or artificial, is
defined by capacities to influence other machines and its capacity to be influenced by other machines. Such power is exploited in the design of laboratory
instruments. The images at the eyepiece are effects of unobservable properties. Although Hooke relied upon causal properties of natural machines in
the design of compound microscopes, he never offered detailed philosophical
arguments on behalf of the reality of such properties. Not surprisingly, his
attention to the mechanical arts drew him away from a rich philosophical
tradition that defined the metaphysics of substance by its causal capacities
to affect change in the empirical world. As I show below, one towering figure
in this tradition is Kant. Although he never examined the rationale for using
instruments, Kant advanced a dynamical conception of matter that is well
suited to serve as the metaphysical basis to Hooke's mechanistic conception
of nature. For Kant, the metaphysics of substance is inseparable from the
means to detect change. Whatever the essence of an unobservable body may
be, it is revealed through its capacities to produce change.

The categories of substance and motion have plagued natural philosophy
since the time of the ancients. For Aristotle, a natural thing exists by virtue of
the fact that it is a source of movement. A principle of change actually resides
in the thing itself, in the sense that a body's metaphysical identity rests on
its causal capacity to generate change: nature is a cause of being moved and
of being at rest within a thing. An apple, by virtue of being a heavy object,
carries within it a principle of falling. Kant retained Aristotle's commitment
to a close metaphysical connection between substance and motion, but he
explicitly dismissed Aristotle's essentialism on the issue of the nature of reality. For Kant, the metaphysics of substance is inseparable from the conditions
for knowing its properties. The only way in which substance can be known
is through its (causal) influence on material bodies: the enduring character
of substance is revealed through its capacities to effect certain changes in a
body's empirical attributes. For Kant, the dynamics of unobservably small
entities are revealed through critical reflection on the conditions for possible
experience, including the motion of appearances from hypothetical entities
(such as atoms) to actual perception.[4]

In his philosophy of science, Kant walked a philosophical tightrope between Locke's liberalism and David Hume's conservatism. On the one hand, Kant demanded a liberalized methodology, which permitted the existence of unobservably small entities, such as atoms and molecules. These entities must be included in our ontological commitments in order to explain the occurrence of empirical events. On the other hand, the dangers of metaphysical speculation, as evident in the Scholastic metaphysics of the medievalists, were lurking in the background (Duncan 1986, p. 279). Hume's work liberated Kant from the need to identify transcendent eternal essences. For Kant, properties of material substances are detectable, even though we are forever shielded from access to the inherent essence of matter. How is this possible? Kant examined the conditions that must arise in order for a scientific investigation of the world to be possible. Such conditions are not found in the thing-in-itself but rather in the categories and principles of cognition, contributing to his Copernican revolution in philosophy. According to Kant, the empirical world is knowable through an objective unity of sensations and conceptions. The possibility of empirical knowledge rests on both the content of sensory appearances and the powers of cognition to construct metaphysical concepts of existence, substance, and cause. Rules for organizing appearances into genuine experience are also rules for identifying a metaphysics of substance.[5]

The need for magnifying instruments was obvious to Kant, but the rationale for their use demanded philosophical attention. When magnifying devices are used, the connection between (hypothetical) unobservable processes and actual perceptions requires a causal sequence of states in space and time. For Kant, knowledge of unobservable processes does not require the immediate perception of actual objects but only the mediated connection between hypothetical entities and some actual perception (Kant 1965, A225/B273). Theoretical entities are linked to perceptions according to known causal laws of science. Such a connection requires a causal sequence of states in space and time (Duncan 1986, pp. 278–79) and is knowable through motion of appearances.

A Kantian substance is detectable through a sequence of appearances, explained through the dynamics of attraction and repulsion, and evaluated through the faculty of judgment. Knowledge of imperceptibly small entities rests on three requirements that establish the relationship between motion and substance.

> (A) Properties of unobservably small entities are detectable through an ordered sequence of appearances which move from hypothetical unobservable entities to actual perceptions.

Every genuine experience requires a causal sequence of appearances. Again, Kant greatly admired Hume's philosophical caution concerning questions about existence, but he could not endorse Humean causation.[6] Hume targeted not only the theologians and Medieval Schoolmen for their wild speculations but also anyone seeking causal powers of nature. Claims about a body's power (capacity, potentiality, capability) were dismissed by Hume as absolute fantasy. But Kant recognized the necessity of positing causal powers as a requirement for the possibility of genuine sensory experience. A material body is knowable through its capacities to influence, and to be influenced by, other material bodies. Every material body, even a single atom, is reducible to its dynamic relations to other bodies. Imperceptibly small entities are knowable through their capacities to effect empirical change. This can be summarized as follows:

(B) The motion of appearances, as described in (A), is generated by the causal capacities of substances to produce changes in other bodies.

Imperceptibly small entities are knowable through their capacities to generate empirical changes, which are sometimes detectable through instruments. Empirical events can occur only after causal powers are exercised. Matter does not fill space by its mere actualized existence but rather by its power to repel and to attract other bodies. The dynamic conception of matter defines each "quantum of matter," to use Kant's phrase, as the power to fill a space through the conflict of attraction and repulsion (Kant 1993, p. 27). Causal powers are manifested when one material body repels or attracts another. Repulsion gives a body its power to resist possible intrusion by another body that may press into "its space" (Kant 1985, p. 498), while attraction, a force of movement of one body toward another body, must be exerted to prevent the universe of bodies from flying away from each other indefinitely. The property of hardness, for example, is an effect of repulsion, anticipating certain developments of nineteenth-century field theories. The entire physical universe is a grand composite of continuous pulsations, comprising repulsions and countervailing attractions.

Often overlooked in Kant scholarship is the purposive character of experience. Not derived from any particular wish or desire, or from an inaccessible spirit beyond our comprehension, purpose is defined by commitment to the goals of scientific inquiry. For Kant, every empirical investigation has a purpose that is driven by the following presupposition: Nature, even in its empirical laws, "adheres to a parsimony suitable for our judgment and a uniformity we can grasp" (Kant 1987, p. 213). A scientist thinks of nature as a system of empirical laws that conforms to his or her understanding, thus establishing a unity between matter and judgment: "[N]ature in its empirical laws harmonizes necessarily with judgment. . . . [This] harmony of nature with our judgment is there merely for the sake of systematizing experience, and so nature's formal purposiveness as regards this harmony can be established as necessary" (p. 233). So, purposiveness is not a subjective desire and has nothing to do with a feeling of pleasure. It is, rather, an objective aspect of our judgments about inquiry (p. 225). The efficient operation of empirical detection operates as a working assumption for all experimental inquiry.

The unity of nature and consciousness through judgment stands at the center of Kant's transcendental philosophy. The mind is engaged in a reciprocal game of a "to and fro" between *actio* and *reactio* (Förster 1991, p. 44). To be possible, experience requires a reciprocity between the so-called physicality of matter and the cognitive powers of mind, with judgment being the "highest" faculty. As the essential power of cognition, judgment unites faculties of consciousness with a system of nature.[7] Through the power of judgment, scientists can inject systematic order into an otherwise haphazard collection of appearances. This leads us to the third of Kant's assertions:

> (C) The causal properties of substance, as described in (B), must be subject to critical evaluation through powers of judgment.

Purpose in this context is not derived from any particular wish, desire, or feeling of pleasure, nor is it drawn from a transcendent spirit. Fleeting human sentiments have no bearing on objective purpose. Purposiveness establishes constraints on scientific investigation based on commitments to certain goals of inquiry (Kant 1987, p. 225).[8] As a precondition of any experiment, nature exhibits a universal order that can be understood through empirical investigation. The entire duality between physical reality and human constructions must be reformulated to reflect laboratory research practices. The harmony between a system of matter and human understanding is guided by the principle of the purposiveness of nature, which functions as a precondition of the scientific investigation of atomic processes, underlying the use of the mediating technologies (Rothbart and Scherer 1997).

Micrographia is one of the great works on the philosophy of experimental inquiry. In detailed descriptions of his experiments, Hooke directs experimenters in the skilled use of microscopes. A mechanistic explanation of experimental inquiry emerges, establishing standards for the interaction between investigator and nature, drawing upon insights from the mechanical arts, mathematics, laboratory research, and the visual arts. Each experiment is itself a kind of mechanistic system that exploits the instrument's power, the specimen's capacities, and the experimenter's skills in operating machines, both artificial and natural. Of course, optical instruments enhance sensory capacities, but such enhancements are explained by the techniques of the mechanical arts, revealing the power of "nature's machines" at the deepest regions. Underlying the optics of vision are the machinists' manipulations of material bodies. Again, the fundamental prototype for experimental inquiry centers on a machinist's hand and eye in the skilled use of machines rather than on an observer's sensory organs in perception.

Four themes can be gleaned from Hooke's microscopy. First, Hooke explicitly endorsed the use of compound microscopes as dynamic tools for revealing the workings of otherwise unobservable machines. Second, suppositions about any natural body can be drawn from analogical associations to the workings of artificial machines. Third, any machine of the natural world is characterized by its capacities to produce movement and to its reagent properties. Fourth, to apprehend a machine's capacities, experimenters must acquire a designer's skills of visualization.

These four themes exhibit striking affinities to aspects of modern spectrographic technique, showing the currency of Hooke's philosophical ideas to the tasks of contemporary researchers, as I argue in chapter 6.

6

ATOMS: EASIER THAN
EVER BEFORE

The surface was invented by the devil!
—Wolfgang Pauli, qtd. in "The Scanning Tunneling Microscope,"
by Gerd Binnig and Heinrich Röhrer

As discussed in the previous chapter, natural philosophers of the seventeenth century believed God's presence was revealed directly in natural machines and indirectly in causal mechanisms deployed for constructing artifacts. For Robert Hooke, a compound microscope operated according to the same principles that govern machines of nature. Indeed, the entire cosmos, he believed, is comprised of small machines embedded in large ones. The empirical landscapes reveal bits and fragments of embedded machines; even the air is composed of tiny springs. Of course, optical instruments were needed to enhance sensory perception. But the sensory capacities exercised during research were deployed for material manipulations, demanding the sincere hand and faithful eye of a machinist. Natural philosophers such as Hooke foresaw the day when the true properties of nature's tiniest machines would be exposed through microscopic studies. The inner mechanism of nature, like the workings of a clock, would be revealed through the use of instruments. He anticipated a time when the bits and fragments of the smallest machines would be observable from advances in instrumentation.

That day has come for nature's machines at an incredibly minute level—the level of atoms, electrons, neutrinos, and quarks. Scientists can detect atoms easier than ever before, according to an advertisement by NanoScience Instruments. With the stunning advances in scanning technologies, atoms have been observable since 1981. The term "observable" means here "experimentally accessible," dependent upon detection methods available to researchers at a particular time. When using modern spectrometers, for example, analytical chemists monitor changes that are generated from technologically induced agitations, manipulations, and interferences. These researchers work with signals, handle information more than observe changes in the specimen's state, and convert signals to readable forms. In this context, then, scanning technologies enable atoms to be "seen" by experimenters.

Such technologies are among the philosophical instruments of the twentieth and twenty-first century. The stunning advances in scanning techniques demand attention to more than the materiality of apparatus. Innovations in instrumentation may require experimenters to adopt a new stance on the conditions for knowledge acquisition. Such a stance is revealed through analysis design plans for instruments. Simply put, a philosophy from instrumentation, not of instrumentation, emerges from a critical examination of design plans.

For philosophers committed to logical empiricism, instruments are needed merely to enhance, or improve upon, the naked senses. Access to a realm of properties at various scales of measurement rests on the capacity of instruments to offer quasi-transparent methods, producing empirical findings that are valid, reliable, and trustworthy. But for today's philosophical instruments of analytical chemistry, the experimental requirements for "valid" signals demand that chemists focus on the following questions, which lead in directions away from logical empiricism: How are signals produced from a particular technique? Did the signals emanate from a specimen's properties or from external sources? What mechanisms were causally responsible for generating the signals? In addressing these questions, the dualities of valid/invalid, reliable/unreliable, and trustworthy/untrustworthy are quite misleading for purposes of securing instrumental data. There is no possibility of acquiring analytical signals in an absolute way, that is, no possibility of removing all the contaminating effects of noise. Research chemists using such techniques are faced with an optimization problem, the solution of which requires maximizing the frequency of desired signals and minimizing the frequency of noisy ones. Such a problem requires more than a mastery of relevant materials. The solution to the optimization problem demands iconic modeling of the causal mechanisms for producing valid signals. Working backward from data entries, experimenters are expected to visualize the generative processes that are responsible for signal production. Through a kind of hypothesis testing, experimenters mentally replicate the generative process through vicarious participation in a thought experiment.

In this chapter, I examine the philosophical underpinnings of a valuable instrumental technique of analytical chemistry—the scanning tunneling microscope. After a brief discussion of the design of analytical instruments in general, I discuss the importance of both noise-reducing and noise-enhancing techniques. Three instruments are then briefly discussed—scanning electron microscopes, field ion microscopes, and scanning tunneling microscopes. Particular attention is given to the U.S. Patent of 1982 submitted by the designers of scanning tunneling microscopy, Gerg Binnig and Heinrich Röhrer.

THE ENGINEERING OF INSTRUMENTS

Instrumental technologies lie at the heart of experimental research in chemistry today. Every analytical instrument is designed to convert an analytical signal that is not directly detectable to a form that is (Skoog and Leary 1992, p. 3). The resources include a signal generator, detector, signal processor, and readout device. Figure 17 lists most of the analytical signals that are currently used for instrumental analysis.

FIGURE 17. Signals employed in instrumental methods

Signal	Instrumental Methods
Emission of radiation	Emission spectroscopy (X-ray, UV, visible, electron, Auger); florescence, phosphorescence, and luminescence (X-ray, UV, and visible)
Absorption of radiation	Spectrophotometry and photometry (X-ray, UV, visible, IR); photoacoustic spectroscopy; nuclear magnetic resonance and electron spin resonance spectroscopy
Scattering of radiation	Turbidimetry; nephelometry; Raman spectroscopy
Refraction of radiation	Refractometry; interferometry
Diffraction of radiation	X-ray and electron diffraction methods
Rotation of radiation	Polarimetry; optical rotary dispersion, circular dichroism
Electrical potential	Potentiometry; chronopotentiometry
Electrical charge	Coulometry
Electrical current	Polarography; amperometry
Electrical resistance	Conductometry
Mass-to-charge ratio	Mass spectrometry

The first six types of signals involve electromagnetic radiation; the next five signals involve changes in a beam of radiation brought about by its passage into the sample. A sequence of signals emanates from the transformation of energy, information, and specimen. Each technique requires manipulation and detection. The manipulation leads to conversions that occur with respect to energy, information, and (pure) specimen. For any testing device, information is received, prepared, and compared or combined, then transmitted, displayed, or recorded. Information can be transferred or stored in data banks, for example (Wallace 1977, p. 22).

The entire system of signals constitutes a model for processes occurring outside the laboratory. The signal's progression, transformation, and response to causal influences simulate real-world processes. Internal influences on the signal are modeled after macro-dynamics of processes that occur generally. The model represents a microscopic universe that evolves according to uniform rules of interaction. This universe simulates global influences.

But the production of signals rests on factors that go beyond electronic circuitry, factors that are not reducible to the dynamical properties of experimental phenomena. The principles of physical sciences alone cannot account for the pivotal distinction between desirable signals and noisy ones. In any experiment, some experimental phenomena carry information about a specimen's properties, but other phenomena are neither sought nor intentionally exploited and can, indeed, have damaging effects on the experiment. The results are noisy signals. To determine a signal's integrity, experimenters must locate various sources of influence, some constructive and others contaminating, in preparation for corrective measures.

The term "noise" refers to the kind of distortion that the system undergoes from an "undesirable" influence, generally from a source that is extrinsic to the intended measurement system. A noisy signal, by definition, has a contaminating effect on the accuracy and precision of an analytic signal by negatively influencing the evolution of the signal within the instrument (Skoog and Leary 1992, p. 46). A desired signal, in contrast, is an information-carrying variable that has its source in a specimen's properties, based on a specimen's physical response to an instrument's probing manipulations. The determination of noise-generating phenomena and phenomena producing desirable signals rests on goal-oriented techniques that the physical sciences alone cannot explain. An instrument's accuracy is never measured in isolation from undesirable interferences, and every measurement has certain ineliminable levels of noise associated with it (Malmstadt and Enke 1963, p. 193).

A noisy signal does not carry information about the specimen but carries "extraneous" information (Coor 1968). For a noisy signal, the sources (its voltage or current) of influence are foreign to the specimen's character. From the perspective of electronic circuitry, there is no difference between a desired signal and noise; both are explained as variations of voltage with time. We might say that they are both signals. Noisy signals are said to degrade the integrity of analytical signals, because the causal influences on the output were not present at the input (Malmstadt and Enke 1963, p. 190). Noise can damage the experiment in three ways: first, a message might be unintelligible; second, noise can cause the receiving system to malfunction, producing meaningless data bits; and third, the efficiency of the communication system can be threatened, requiring, for example, extra transmitting power to the original signal (Schweber 1991, pp. 64–65).

Some sources of noise are external to the progression of the analytical signal. The integrity of an analytical signal can be corrupted by the wiring electricity circulating around the walls of the laboratory and building in which the research is performed. Additionally, external sources of noise may originate from almost any electrical device in close proximity, because such sources emit electromagnetic waves as a byproduct of their operation. Some of these sources include AC power lines, radio and TV stations, engine ignition systems, and lightning. Atmospheric conditions, such as thunderstorms, may emit radiation of up to 30 MHz. The sun is a source of low-level noise that begins at approximately 8 MHz (Currell 1987, pp. 179–80). The conductor in an instrument can function as an antenna capable of picking up electromagnetic radiation and converting it to an electrical signal (Skoog and Leary 1992, p. 49).

Not all sources of noise are external to the measurement system itself. Internal sources of noise arise from any power-dissipating device. Of course, every laboratory apparatus is a potential source of contamination. For example, Johnson noise arises because the electrons that carry an electric current always have a thermal motion, causing small fluctuating voltages across any resistor in the electrical circuit. This noise is often called "white noise" because (like white light) it contains all the possible frequency values (Currell 1987, p. 180).

In absorption spectroscopy, deviations from Beer's law in actual research are, in some cases, sources of noise. Beer's law holds that the ratio of the powers of transmitted and incident radiation are proportional to both the mass absorption coefficient and the density of the sample. So, the absorbance A of a solution can be defined as the ratio of energy of radiation Po impinging on the sample solution over the energy of radiation P that is transmitted. So,

$A = log Po/P$

The actual absorbance depends on the captured cross-section of the species and on the probability for an energy-absorbing transition occurring. The absorbance A of radiation by a concentration is directly proportional to the path length b through the solution and the concentration c of the absorbing species. The absorbance of radiation is idealized by the equation

$A = abc$

where a is the absorptivity of the solute, as determined by previous experiments. But the experimental absorbance that is measured in the laboratory invariably deviates from the "true" absorption as defined by Beer's law. These deviations are explained by interference from a variety of possible sources. For example, as the radiation beam passes through a transparent container holding the sample solution, the power of the beam is decreased significantly. This is caused by reflection at the two air/wall interfaces as well as at the two wall/solution interfaces, as shown in Figure 18 (Skoog and Leary 1992, p. 125). This results in the substantial attenuation of the radiation beam, deviating from the absorbance predicted from Beer's law.

The damaging effects of noise are never completely removed from the scene. At best, the total noise that is identified can be minimized and the analytical signal maximized; one can never measure an analytical signal in an absolute sense. The pivotal factor in determining the accuracy of the instru-

FIGURE 18. Reflection and scattering losses as radiation passes through a transparent container

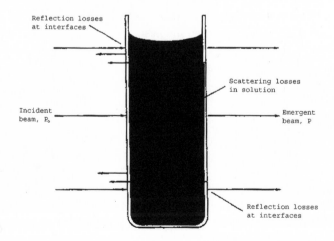

mental data is the signal-to-noise ratio. The integrity of specific measurements is a function of optimizing this ratio: S / N, where $0 < N < 1$ and $0 < S < 1$.

One might conclude that the optimization problem above leads to an infinite regress, because any determination of S must be measured against the contaminating influences of noise N. But such skepticism is unwarranted. Although a noise-free laboratory is unattainable, chemists have devised techniques to minimize or compensate for the effects of noise. Judgments regarding a signal's worth are achievable through a kind of hypothesis-testing about its source. The signal's evolution is compared against an idealized conception of an experiment, determined in large measure by the instrumental design.

Environmental interference can be reduced by blocking, grounding, or decreasing the lengths of conductors. One technique involves electrostatic shielding, in which some of the instrument's wires are surrounded by a conducting material. Shielding is designed to minimize the effects of uninvited intrusion that arise from the instrument circuitry. Electromagnetic radiation is absorbed by the shield rather than by the enclosed conductors. The conductors are wrapped with aluminum or copper foil, with a drain wire in electrical contact with the foil, providing almost perfect covering. But other sources of noise cannot be eliminated. Thermal noise is a voltage having its origin in thermally induced motion of charge carriers (Malmstadt and Enke 1963, p. 193).

TURNING NOISE UP

The presence of noise shows how certain manipulating agents become harmful contaminants, eroding the otherwise efficient production of information. In his important contributions to the development of high resolution NMR spectrometers, Hans Primas used stochastic methods to reveal beneficial effects of noise for characterizing physical systems, demonstrating the value of noisy signals for nonlinear physical systems in chemistry. Because the NMR signals were invariably weak and noisy, Primas was intrigued by the stochastic processes associated with electronic measurements and data transmission. In an important breakthrough in NMR spectroscopy, he discovered that the detection of noise could have beneficial effects on physical systems. In effect, such "noise" ceases to be noise when exploited constructively for purposes of research.

Transferring certain techniques from the conventional domain of telecommunications and control engineering to an entirely different domain of chemistry, Primas established new applications for physical chemistry. He discovered the great value of random noise for identifying nonlinear physical characteristics of system in chemistry. Although this process of noise detection is used to separate signals from noise, the use of the Wiener-Khintchine theorem to describe an atomic system as a stochastic process was highly innovative. From the perspective of this theorem's application, not only does the signal provide information about a specimen's properties, but the noise does as well. Indeed, Primas provided the first description of the response of a nuclear spin system to a stochastic perturbation based on the filtration of signals from noise.[1]

TECHNIQUES OF MICROSCOPY

In this section, I examine scanning tunneling microscopy (STM), a powerful technique that can be deployed in air, inert gas atmospheres, and liquids.[2] Because of its versatility and reliability, STM has produced stunning results related to, for example, the surface of silicon, the distinctive coils of DNA, and the biochemical processes associated with blood clots. Below, I briefly address problems arising from the use of scanning electron microscopes and field ion microscopes, followed by an exploration of the design plans for STM. With their discovery of STM, Binnig and Röhrer circumvented otherwise intractable problems associated with two earlier methods.

Scanning Electron Microscopy (SEM)

A forerunner to the imaging techniques can be found in the X-ray diffraction method, developed in 1912 by Max von Laue. After obtaining the first X-ray diffraction pattern, van Laue's success encouraged Lawrence Bragg to study alkali halides with such techniques. Yet Bragg recognized the problems with deploying this method. In particular, organic molecules are much more difficult for analysis of their structure than are inorganic ionic lattices because of the large number of parameters needed to define the structure of organic molecules (Morris and Travis 2002, p. 62). Nevertheless, this method provided researchers with an excellent means for studying the arrangement and spacing of atoms in crystalline materials. The resulting knowledge of physical properties of metals and polymeric materials led to major industrial applications. X-ray diffraction studies also offer an invaluable resource for understanding the structures of complex natural products such as steroids, vitamins, and antibiotics (Skoog and Leary 1992, p. 378). For example, X-ray diffraction studies of the structure of vitamin B12 by Dorothy Crowfoot Hodgkin represented an outstanding breakthrough in organic chemistry. In her publication of the full structure of this molecule in 1957, she provided for the first time a structural elucidation of a complex molecule almost entirely by physical methods.

Familiar to researchers in chemistry, material science, geology, and biology, SEM provides detailed information about surfaces of solids on a submicrometer scale. The technique occurs as follows: After surface atoms of a metal are bombarded with high-energy electrons, the spatial distribution of electrons is measured. To obtain an image, the surface of a solid sample is swept in a raster pattern. A raster is a scanning pattern in which an electron beam (1) sweeps across a surface in a straight line, (2) returns to its starting position, and (3) shifts downward by a standard increment (Skoog and Leary 1992, p. 394).

This technique solved a major problem associated with image resolution. In performing an electron microprobe analysis, the spatial resolution of the image is limited by the fact that the probing particles must be smaller than the dimension of the atom structure being observed. As an analogy, imagine trying to detect scratches on a smooth surface by running a finger across the surface. Those scratches that are much smaller than the dimension of the fingernail would go undetected, while the scratches that are larger than the width of the fingernail would be easily felt. In SEM, the problem of spatial resolution arises from the fact that a typical wavelength of visible light is 600 nanometers (nm) wide, and the approximate diameter of an atom is 0.2 nm. Since the wavelength of visible light is several thousand times larger than the width of an individual atom, the placement of individual atoms is unobservable. To solve this problem, designers developed a technique in which the wavelength of the probing electrons is smaller than the atomic size of 0.2 nm (Bonnel 1993, p. 436).

Despite this important advance, designers quickly identified three significant deficiencies with SEM. First, the images of a surface can be produced only in a vacuum, which is difficult to produce; second, the observed images may not accurately represent the original surface structure; and third, electrons with high kinetic energy penetrate relatively far into the surface and are then reflected back to the detector. This process of penetration and reflection distorts the image of the top one or two atomic layers. Because of such distortion, the best result that researchers can expect is an average of many atomic layers. To overcome these deficiencies, scientists explored alternative methods for detecting atomic surface properties.

Field Ion Microscopy (FIM)

The first technique for detecting properties of individual atoms on metallic surfaces was developed by Erwin Müller in 1936 (Chen 1993, p. 412).[3] This instrument, known as a field ion microscope, has a simple design comprising a vacuum system, a needle tip, and a phosphorescent screen. Müller's diagrams and discussion enable readers to visualize the technique he developed (Müller and Tsong 1969, p. 99).

As shown by Figure 19, high voltage is applied between the filament and screen. After positioning the needle tip on a specimen, helium or neon atoms are adsorbed directly on the atoms under investigation. A potential on the order of 10,000 volts is then applied to the surface, causing helium or neon atoms to be released in the form of ions and directed toward the fluorescent screen. On their way to the fluorescent screen, the distances between neighboring ions increase, causing a magnification of the distribution of atoms on the surface. This process is shown in Figure 20 (Müller and Tsong 1969, p. 99).

FIGURE 19. Original field ion microscope

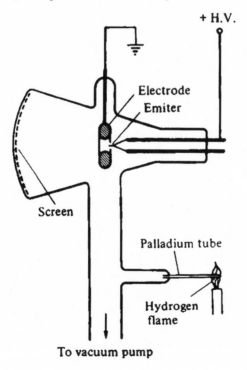

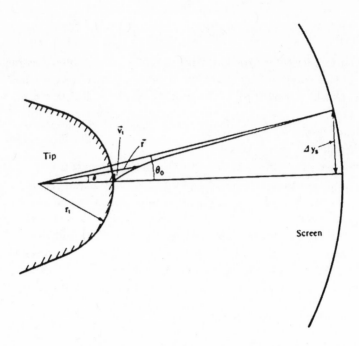

FIGURE 20. Representation of the path of an ion from the tip to the screen of a field ion microscope

The atoms are shown to move farther away from one another after they hit the screen than when they lie dormant on a surface. As the helium or neon atoms hit the screen and produce an image on the surface structure, they retain their original relative positions.

To persuade the scientific community, designers of FIM constructed a physical model of a tungsten tip based on analogies to a hemispherical ball, as shown in Figure 21 (Müller and Tsong 1969, p. 78).

By painting the kinks (protruding edges of individual planes) with a phosphorescent coating in the model, the pattern of light and darkness was observed to be similar to images obtained in FIM, as shown in the right diagram of Figure 21. The physical model replicates the pattern of images produced when xenon atoms are released from the kinks. Although this model helped to explain observations of such images, the model was difficult to construct due to the large number of atoms (balls) that were needed.

As with SEM, experimenters soon identified deficiencies in FIM. Like electron microscopy, FIM requires that images of a surface be produced in a vacuum. Additionally, the image area is limited to about 0.8 nm^2. Further-

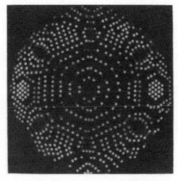

FIGURE 21. *Left:* Ball model for an ordered surface. *Right:* Ball model with kink site atoms painted with phosphorescent paint, then viewed in the dark

more, low temperatures are required for operating the microscope; neon and helium will not absorb on most surfaces except at either liquid nitrogen or liquid hydrogen temperatures.

Scanning Tunneling Microscopy (STM)

The properties of surfaces have provoked fascination and frustration among chemical researchers. Binnig and Röhrer, both working for IBM in Zurich, discovered a means for producing high-resolution measurements that offered the following improvements over the topographic technique: (1) the distance between emitter tip and sample was decreased; (2) the required voltage between the two was lowered; (3) the emitter tip was sharpened; and (4) vibrational dampening was added (Binnig et al. 1982). For their discovery, they won the 1983 Nobel Prize in chemistry. Invoking the famous curse of Wolfgang Pauli, "The surface was invented by the devil!" (Binnig and Röhrer 1985, p. 50), Binnig and Röhrer devised a technique from major advances in quantum tunneling. Readers of their design plans are invited to visualize how a tunneling current is deployed to "sense" the atomic and electronic surface structure of a compound. The design plans are tested through a thought experiment in which one anticipates how STM could be used in research, what events at the microscopic level would occur, and how the major challenges associated with other techniques of surface chemistry can be overcome. In this respect, the reader becomes a virtual witness.

Through STM, a sharp needle is brought within a few angstroms of the surface of a conducting sample; the needle sweeps across the surface, moving up and down, left and right, forward and back, in increments of one angstrom. A tunneling current detects features of a compound's surface structure. Of course, a robust understanding of this technique demands attention to the major challenges associated with surface chemistry, challenges that seem to arise from a netherworld of demonic powers.

Binnig and Röhrer offer a detailed conception of this technique in their 1982 patent filed with the United States Patent Office (Binnig and Röhrer 1982). From this patent, the instrument's components are depicted in Figure 22.

The apparatus includes a metal tip (indicated by 5 in the figure), a probed sample surface (4), and a chamber (1) for generating an ultra-high vacuum. The chamber contains apparatus for supplying energy and various controls needed for scanning. The instrument is defined through both static models of the apparatus and dynamic models of electron tunneling. In STM, as in the other techniques described, a needle rides across a prepared surface. By passing a current from the needle to the surface of the material under examination, one is able to "feel" the topography of the surface. The voltage applied to each from the control electronics determines the exact position of the x, y, and z directions. Special circuitry is used to measure the current between the tip and the sample. Computers are deployed to manage the control signals to the electronics as well as the signal from the tunneling current. In practical terms, a topographical experiment begins by positioning the tip properly in relation to a sample. The tip is then moved in the z-direction to adjust the sample-to-tip distance, which must not exceed a few tenths of a nanometer. A small bias voltage is then applied between the sample and the tip. As the tip approaches the surface, a tunneling current guides the final

FIGURE 22. Scanning tunneling microscope

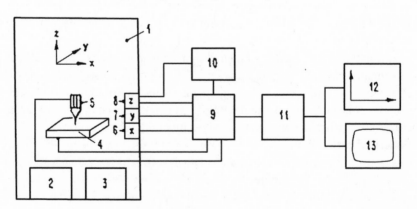

positioning in the z-direction. The X-scan voltage is then ramped, causing the tip to move in the x-direction but not in the y-direction.

The vacuum tunnel effect is utilized with this microscope. In an ultra-high vacuum at cryogenic temperature, a fine tip is raster-scanned across the surface of a conducting sample at a distance of a few angstroms. The vertical separation between the tip and sample surface is automatically controlled as to maintain a constant value proportional to the tunnel resistance, such as tunneling current. The position of the tip with respect to the surface is controlled preferably by means of piezoelectric drive acting in three coordinate directions. The spatial coordinates of the scanning tip are graphically displayed by showing the drive current or voltages of piezoelectric drives.

In Figure 23, the roughness of the surface is shown in the dotted line, indicating the path of the tip at a distance over the sample surface. The dashed line represents the path of the shadow of the tip over the surface. The tip (5) is poised above the sample surface by a distance corresponding to the vacuum tunnel barrier, approximately 2 nm. Scanning may occur in any direction. The processes of vertical and horizontal scanning should be adjustable over small distances in all directions, as indicated in Figure 24.

In one sweeping motion, the tip moves laterally along the x-dimension, as indicated by (19) in Figure 24. This is followed by another sweeping along the thickness of the raster line in the y-dimension. The tip is then shifted in the direction for a vertical scanning in the z-dimension, indicated by the double arrow (20). Vertical scanning is particularly hazardous, because the

FIGURE 23. Illustration of needle riding over rough surface

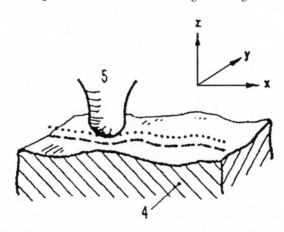

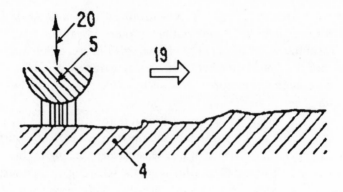

FIGURE 24. Tip moving both vertically and horizontally

scanning must be very finely tuned with an accuracy on the order of fractions of an angstrom.

In STM, the probing electrons escape interaction with the surface atoms by traveling on a different route than they "normally" would, thereby avoiding experimental interferences. Under normal conditions, a relatively large amount of energy (a few eV) is required to remove an electron from an atom. As shown in Figure 25, an electron can be freed from the influences of a surface atom only by acquiring an amount of energy that surpasses the so-called energy barrier.

The energy barrier is equal to the difference between the charge of an electron attached to the surface atom and the charge of an electron that is freed from the influence of a surface atom. But the amount of energy needed to overcome the energy barrier is impractical to produce under laboratory conditions.

Quantum tunneling allows an electron to travel from one atom (on the emitter tip) to another atom (on the surface) by a low-energy route. This route is made possible due to a small overlap of the wave functions of each electron (Atkins 1978, p. 402). From the perspective of energetics, the electron travels to a surface atom by tunneling through, but not over, the energy barrier. Tunneling occurs when electrons are freed from the potential barrier without the thermal energy that is normally required. In their 1982 patent, Binnig and Röhrer show how tunneling can occur between two metals (1982, Fig. 2).

In Figure 26, the electrons of the first metal are bound in a first potential well, as indicated by (14) in the figure. A second metal is characterized by a second potential well (15), which is separated from the first potential well by a potential barrier (16). For tunneling to occur, electrons must leave the first

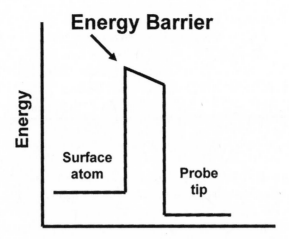

FIGURE 25. Energy barrier when an electron travels from a surface atom to the probe tip of an STM

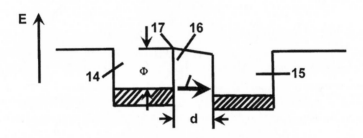

FIGURE 26. Tunneling of electrons between two metals

metal and enter into a vacuum when their energy (E) is raised to the value of the upper limit of the bound energy states, the so-called emission edge (17) (Binnig and Röhrer 1982, p. 6).

WITNESSING THE SOURCE

Of course, the instrument may be poorly designed, the apparatus may malfunction, and the researchers may be inept. However, the possibility of signal contamination alone does not guarantee that errors will occur. Experimenters have fine-tuned specific skills to minimize such dangers. A skeptical experimenter will show signs of neurosis if the channel conditions are checked beyond what is necessary (Dretske 1981, pp. 115–16). The experimenter's ability to acquire technological power cannot be explained entirely by sociological investigations of community practices and the vested interests of researchers, nor can it account entirely for the results of experiments. A deeper study of the interaction between instrumental techniques and real-world processes is indispensable for a robust account of such power.

The assessment of signal-to-noise ratios rests on explanatory models of the causal mechanisms at work. The production of noise is a major topic of virtual witnessing. But noise is not always harmful to the system, as illustrated above in the innovative techniques discovered by Primas from his contributions to NMR spectroscopy. In experimental settings, the laboratory simulations of global processes include both sources of contamination and sources of compensatory mechanisms to override such contaminants. Sometimes, nature acts as invited guest. But when nature acts as unwanted intruder, experimenters are expected to deploy techniques to minimize noisy signals. With such modeling techniques, optimization problems must be solved in order to convert signals (both desirable and noisy) to readable information.

In their causal explanation for tunneling, Binnig and Röhrer resort to iconic models of both particles and waves. A reader of their patent is expected to imagine electrons moving through, and not over, an energy "wall," as if witnessing movement through the potential energy barrier. Binnig and Röhrer use the image of particles moving through walls as an icon for the behavior of electrons. Electrons surpass an energy barrier, which is analogized to a solid wall, only if they have enough energy. Tunneling occurs when the distance between the surface and tip atoms of the microscope is small and when a small voltage is applied between the surface and tip (Binnig et al. 1982). The mechanics of both waves and particles are conveyed in the following passage:

> In a[n] atomic system or in a solid body, if charged particles are subjected to an interaction composed of a long range repelling component and a short range attractive component, then the resulting force builds to a potential wall or a barrier. According to classical conceptions, such a barrier can be crossed only by particles having energy greater than the barrier. There are nevertheless always a finite number of particles by a potential barrier, which are capable of crossing the potential barrier although they do not have sufficient energy. In a sense, they undercross the potential barrier via a tunnel. This so-called tunnel effect can be explained only by wave mechanics. . . . According to the tunnel effect there exists a calculable probability that a finite number of electrons bound by a potential can cross the tunnel barrier even at low voltage differences. . . . Some bound

electrons are capable of tunneling through [barriers]. (Binnig and Röhrer 1982, pp. 1–2)

Based on quantum mechanics, an electron's location is characterized as a probability function for entering a classically forbidden region, as if to represent how the electron is "smeared out" in a region between the tip and sample surface (Binnig and Röhrer 1985, p. 53).

Thus, in analytical chemistry, the old dream of observing an arrangement of atoms at the surface of a compound is now realized with the discovery of sophisticated scanning techniques. Approximately 3,000 publications each year can be directly credited to new methods of scanning. Additionally, chemists contribute significantly to standards of research through designs of intelligence systems. In certain respects, chemists perform the work of knowledge engineers—agents who develop or improve upon techniques for extracting information from the environment. The production of artifacts through the manipulation of a sample is essential to such techniques. As knowledge engineers, chemists determine what must be done to generate valid signals, what kinds of interference should be used to probe a sample's properties, or what kinds of results should be expected.

Binnig and Röhrer not only exploited certain portions of the world for the purpose of constructing a scanning tunneling microscope but also justified a new technique by appropriating iconic models from both wave mechanics and particle mechanics. Reliance on the electrons' capacities to tunnel through an energy barrier is critical. Inviting readers of design plans to imagine electron tunneling and engaging them in what I call virtual witnessing, Binnig and Röhrer implicitly illustrate an important aspect of all philosophical instruments—the inseparability of technology and ontology. The technology is defined through the iconic models of those portions of the world needed for information retrieval; modeling, in turn, rests on ontological commitments to real-world processes, which is a topic of primary concern in chapter 7.

7

SPECIMENS AS MACHINES

The mechanistic underpinnings of past centuries have not been expunged from laboratories today. A specimen's "objective" character is deciphered through its capacity to generate new products under specific laboratory conditions. During research, a specimen is characterized by its agentic properties as a physical/chemical source of signals, responsive to dynamic interventions that are technologically induced and in part causally responsible for the information retrieved during research. Measuring devices of laboratory research exemplify true philosophical instruments.

In this chapter, I examine how the nature-as-machine metaphor that captivated scientists for centuries retains its influence today.[1] I reveal how a specimen functions as a tool, mimicking artificial machinery, behaving like artifacts that exist outside of the laboratory. The entities "in captivity" simulate "nature in the wild" (Harré 1998). The use of this metaphor and the comparisons it generates offer insight into tremendous power that instruments give to experimenters. A specimen's capacities and limitations are revealed through the new entities (states, events, or products) brought into existence during an experiment. Such capacities presuppose the workings of nomological machines, to borrow Nancy Cartwright's notion. *In an attempt to compensate for the widening "distance" between experimenter and specimen, designers presuppose that the portion of the world under investigation is endowed with the same kinds of capacities that can be attributed to their own creations.* As the objects examined with today's philosophical instruments become increasingly remote, experimenters move back and forth between the immediate action of the laboratory and other environments, drawing elements of the "outside" world into the laboratory (Gooding 1989, pp. 183–84). Slipping easily between the outside world and the world of laboratory entities, experimenters monitor the dynamics of nature that are found locally but operate globally. Although local, the micro-universe conforms to "global" principles.[2]

The revolutionary advances in biochemistry of the 1950s perpetuated the allure of organism-as-machine (Bechtel and Richardson 1993; Brandon 1985; Burian 1996). A defining mission of biological researchers has been discovery of an organism's productive processes (causal mechanisms) that are responsible for organic phenomena. In their provocative article entitled "Thinking about Mechanisms," Peter Machamer, Lindley Darden, and Carl F. Craver examine the nature of mechanisms in neurobiology (2000). They advance a dualistic conception of a mechanism in terms of entities and activities (p. 3). All biological phenomena are products of the organization of entities and activities. In DNA replication, for example, the DNA double helix unwinds, exposing slightly charged bases to which complementary bases bond, eventually yielding two duplicate helices. In this case, protein synthesis is a mechanism comprising a process (activity) of the hydrogen bonding of a DNA base with a complementary base (entities with their properties) (p. 3).

Although Machamer, Darden, and Craver offer compelling arguments for the centrality of causal mechanisms in the biological sciences, their dualistic conception of a mechanism is unconvincing. Underlying the presence of entities and activities are the specimen's capacities, which are productive properties causally responsible for an organism's state. The use of apparatus induces nature to display its properties, as if exposing them to inspection. The ontological ties that bind causal events with effects are the capacities of particular entities that emanate from a generative mechanism (Harré and Madden 1977, p. 11). Scientists today freely speak of an entity's capacities (such as flexibility, acidity, solubility) to behave in certain ways. The flexibility of a piece of steel means that the steel will exhibit a change of shape from sufficient pressure. Flexibility is defined as a set of conditional manifestations of the entity to exhibit a certain range of behavior. Any tendency of a specimen is revealed empirically under the appropriate releasing conditions and the suppression of obstructing influences. Of course, the steel does not lose its capacity for flexibility when appropriate conditions are not met. Rom Harré writes: "A *particular Being* has a *Tendency* which if *Released,* in a certain type of situation, is manifested in some observable *Action* but when Blocked has no observable effect" (1986, p. 284; emphasis in original).

To be sure, Machamer, Darden, and Craver are well aware of the capacities of a biomolecular entity to act as a cause when engaged in productive activity. But they fail to recognize the centrality of the causal capacities of mechanisms.

The moon has a power over the tides on earth; the earth has a power of attraction over falling bodies; at the molecular level heat has a power to break the hydrogen bond between two water molecules resulting in water vapor; a chemical atom can be understood by its electronic constitution of attracting and repelling forces. Such capacities of these elements are tendencies, which are activated by creating the proper conditions and suppressing potential blocking agents. A cannon ball has a tendency to fall toward the earth, just as an electron has a tendency to accelerate toward a positively charged plate. Such capacities are properties of nomological machines.

For Cartwright, the physical world is a world of such machines (1999, p. 52). A machine's capacities give rise to the kind of regular behavior that we represent in scientific laws (p. 49). In classical mechanics, attraction, repulsion, resistance, pressure, and stress are capacities that are exercised when a machine is running properly. Each law holds true due to the capacities of each machine in the system. Regularity is maintained when the machine runs properly (p. 59). In Newton's law of universal gravitation, for example, the term "force" does not refer to yet another abstraction, like mass or distance. "Force" refers to a capacity of one body to move another toward it, manifested in different settings to produce different kinds of motions. In the physical sciences, the causal capacities of a nomological machine can be conveyed through the dynamics of energy transfer. Energy is a physical quantity that flows from one object to another in various forms, such as heat or electromagnetic energy. The sun causes the earth to warm, since the energy of fusion transferred by photons becomes the thermal energy of terrestrial objects. In photosynthesis, solar energy takes the form of chemical energy. An energy flow is a defining (physical) property of a causal connection between pairs of events (Fair 1979, pp. 232–36).[3]

CONFORMATIONAL FLEXIBILITY OF DNA

Nomological machines are revealed in the dynamics of organisms. The study of organic processes was advanced in the important work of nineteenth-century biologist D'Arcy Wentworth Thompson. He urged biologists to study physical and chemical processes determining an organism's substructure on the grounds that the organism's form is identified by its causal mechanisms. Although most biologists during his lifetime dismissed his proposals, current findings in biomechanics have confirmed some of Thompson's insights. Such mechanisms are conveyed through diagrams of force, which represent physical processes of morphology.[4]

Empirical studies show that molecules are not static, not frozen in a fixed configuration, but spinning and vibrating, as well as repelling and uniting with other molecules. Underlying such activity are internal changes of the compound's dynamic state. Instead of thinking in terms of a single, average DNA configuration, biochemists use dynamic models to convey fluctuations of geometrical configuration. The nucleic acid double helix is often modeled in terms of constituent bases. Variations of local structure in double-helical DNA are expressed at the macro-molecular level. The local flexibility of DNA has been idealized in terms of symmetric bending and twisting models. In some studies, DNA is described through the spatial fluctuations of successive base pairs.

Conformational flexibility has important biochemical implications, playing a role in modulating the biological activity of genes. Through its bending and twisting motions at the level of neighboring base pairs, the chemical sequence generates a higher-order folding of the double helix and facilitates specific binding of proteins. For some biomolecules, the natural curvature in the double helix brings the otherwise distant parts of the nucleotide sequence into close contact. This looping of DNA assumes a role in transcriptional control as well as in other processes involving DNA metabolism (Olson et al. 1993, p. 531).

Flexibility from one base pair to the next is typically described in terms of four degrees of freedom: helical twist, tilt, roll, and propeller twist. Figure 27 depicts the three variables of helical twist, tilt, and roll (Dickerson, Drew, and Conner 1981, p. 6).

Helical twist is the rotation of a base pair around the axis of the double helix, and the axis is normal to the base plane. The positive twist refers to right-handed helix. The bending of a helical structure around its long axis is called roll; the roll is positive if the major groove is compressed. Tilt is rotation around the short axis. Propeller twist is the degree of supercoiling of the entire double helix itself (Sarai et al. 1989, p. 7843).

Elasticity models offer insight into the reasons for such fluctuations. From such models, the positions of atoms in a molecule fluctuate continuously as a result of two types of vibrations—bending and stretching. Bending vibrations occur from some type of a change in the angle between two bonds. The three properties of flexibility in a section of double-stranded DNA (Figure 27)—that is, helical twist, tilt, and roll—are kinds of bending. The fourth property of flexibility, that is, propeller twist, is characterized as vibrational stretching. Consequently, the total fluctuation of DNA is characterized by the types of vibrating motions associated with elasticity models of mechanics.

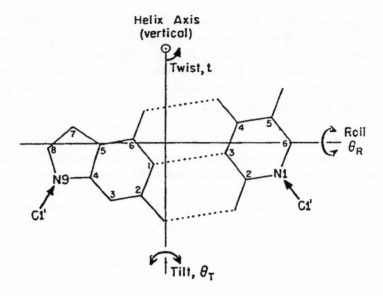

FIGURE 27. Flexibility between base pairs in double-stranded DNA: helical twist, tilt, and roll

Empirical studies have demonstrated the range of fluctuations of the molecule of DNA.[5] The information about base-base flexibility can be guided by rules and resources for constructing energy maps. The conformational flexibility of the double helix can be attributed to the dynamics of energy exchange. The local flexibility is conveyed through maps of energy of free base pairs.[6] Energy maps are contour diagrams of the potential energy associated with the rolling motion (<r>) and tilting motion (<t>) of some base pair. Figure 28 displays the rolling and tilting motion of *AT*, assuming that the angular twist is fixed at 36 degrees (Maroun and Olson 1988, p. 568). Contours are drawn by interpolation of energy at 1 kcal/mole intervals from 1 to 5 kcal/mole. The lowest energy state is indicated by *x* on the map.

The base pair *AT* exhibits a relatively high degree of flexibility because of the absence of an NH_2 group at C2 in the minor groove. In one study, adenine and thymine were substituted for each other in order to elucidate the nature of the three dinucleotide junctions *AA*, *AT*, and *TA*. These junctions are known to produce different effects within DNA fragments. The results reveal considerable variation in the twist angle for these dinucleotide junctions. Moreover, the *TA* junction is distinctly more flexible than the other two

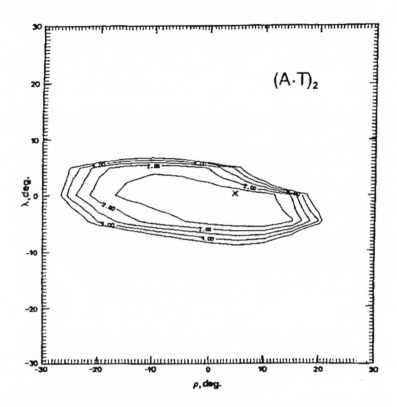

FIGURE 28. Potential energy map for an AT base pair

and is responsible for absorbing the structure stress imposed by overwinding (Zakrzewska 1992, p. 692).

Figure 29 is also a contour diagram of rolling and tilting, in this case for the base pair *GC* (Maroun and Olson 1988, p. 569).

The maps for both *AT* and *GC* reveal that the molecules roll more easily around the long axes than tilt about their short axes. This is explained by the increase of electrostatic and van der Waals interactions as the rolling motions increase beyond zero degrees (Srinivasan et al. 1987, pp. 461–66).

So, the dynamic properties of such a machine are revealed in energy maps of free base pairs, as shown above. The conformational flexibility of the double helix can be explained through the workings of a nomological machine, at least for research purposes. But how exactly can an experimenter achieve access to a specimen's pure form that presumably transcends the demands

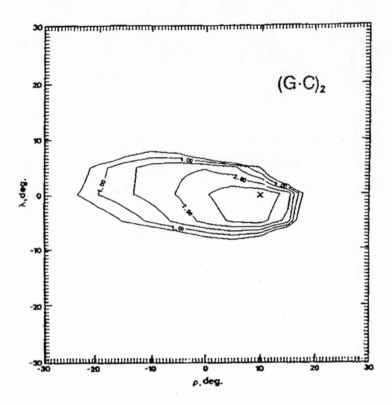

FIGURE 29. Potential energy map for a GC base pair

and limitations of laboratory research? I address this question by exploring the complexity of the notion of a pure specimen.

PURITY OF FORM

In contemporary research, evidence of a specimen's properties is conveyed through a sequence of analytical signals, explained (in part) through the dynamics of experimental phenomena and governed by pragmatic factors associated with the instrument's efficacy. This function of a specimen is presupposed in the instrument's design. Again, a pure specimen functions as a signal-producing machine, characterized by capacities for coding, transmitting, receiving, and decoding information. As a signal source, a specimen is responsive to technologically induced interventions and causally responsible, in part, for the information retrieved during research.

Bruno Latour writes that reality is defined in the laboratory by its capacity to resist intrusion by external forces (1987, p. 93). I prefer to say that reality is defined by capacities that *permit* intrusion from external forces and that are activated in ways that generate detectable effects. A pure specimen is by its function subject to interference and intrusion by some external radiation and definable through its reactive capacities. The notion of a reactive system is a prototype for characterizing the specimen's purity. A pure specimen is not passive; it reacts to technologically induced agitations. Chemical and physical changes occur within a reactive system due to forces such as the phase in which a reaction takes place, the nature of the solvent, the temperature, or the incidence of radiation (Suckling, Suckling, and Suckling 1978, pp. 65–68).

American pragmatists give priority to purposive interventions, relying on the notion of goal-directed action as a fundamental component of knowledge acquisition. According to this view, action is not simply a sequence of behaviors, immune to influences of thought and purpose. The adequacy of a theory is inseparable from the manipulative efficacy of techniques deployed in theory-testing. The current techniques of laboratory research are defined by the revolutionary advances in instrumental technologies of the twentieth century. The notion of signal processing, essential to such technologies, functions as a prototype for methods of empirical research.

One might object that the essential nature of a chemical substance per se carries no signal and thus conveys no information. However, the functionality of a specimen is revealed in the techniques applied to determine its purity. The ability of experimenters to identify a specimen's purity becomes a prerequisite for experimental investigations of microscopic processes. Through purification techniques of chemistry, for example, a sample may be subjected to high temperatures, dissolved from the use of hot mineral acids, or converted to an aqueous solution. Like an ecological system, the sample is cleansed of extraneous elements, and impurities are cast aside through "sterilizing" methods. In some experiments, a pure sample is one whose temperature remains constant in the course of further phase transitions. X-ray diffraction is a valuable technique for determining the qualitative properties of a crystalline material. To prepare the fibers of DNA for X-ray diffraction, the fiber is pulled from a concentrated solution, stretched in order to produce approximate alignment of the long helical molecules with the fiber axis, and then subjected to X-ray beams in front of a photographic film.[7] As radiation passes through a repeating structure, the scattered waves interfere with one another, resulting in a diffraction pattern characteristic of the material. The diffraction pattern consists of spots or short arcs, which show how the atoms are arranged within each unit cell. The positions of all the atoms in the unit cell can be determined by electron density distribution (Mathews and van Holde 1996, pp. 124–26).

Techniques for removing impurities from a sample presuppose ontological commitments. An experimenter is constantly comparing laboratory results to an idealized "picture" of the system's "true" properties. Certain elements of the experimental landscape are sought, others avoided, and still others ignored; known similarities are made salient, dissimilarities are suppressed, and unknown relations are explored. Purification exposes a specimen's natural properties, just as a domesticated animal retains certain instinctive characteristics, to borrow an analogy from Harré (1998). In preparing a sample for instrumental analysis, an experimenter isolates and displays certain attributes of the compound. In contemporary spectroscopy, a pure specimen is revealed through its invariance to purification techniques, commonly associated with variations of thermodynamic conditions. A specimen is dissected, manipulated physically, heated, cooled, separated, or synthesized. The invariance to such an intervention allows experimenters to isolate chemical species as reaction products afterward. The product of such procedures, in turn, can function as a component for further chemical investigation under new controlled conditions (Schummer 1998).

So, an instrument's efficiency, reliability, and accuracy draw attention to nature's productive capacities, that is, its nomological character. When using instruments, experimenters provoke a response from nature by manipulating certain "preexisting" properties. A specimen's detectable properties are revealed in the new entities (states, events, or products) that are revealed through technological inducements. In the context of laboratory research, a specimen's physical structure rests on evidence extracted from its functionality during research, and a specimen's function merges with detection methods used to explore it. A specimen acts as an agent for change, as well as a reagent subjected to influences by other bodies. In this respect, a specimen is knowable by its capacities to produce, create, and generate detectable events. Again, the inseparable of ontology and techné is revealed through a reflection on instrumentation.

AFTERWORD

The philosophical instruments of the seventeenth century sparked a scientific revolution. According to their designers, such devices unveiled truths of nature because they conformed to global principles of a universe that God created and human experimenters replicated. Advances in microscopy by Robert Hooke, for example, demanded commitment to a mechanistic philosophy that defined the kinds of properties that are amenable to instrumental detection. Today's philosophical instruments extend the reach of laboratory researchers through greatly enhanced physical powers. The spectrometers of analytical chemistry are not simply composites of steel, plastic, or other material. Today's instruments are channels of abstract ideas, offering guidance to experimenters, establishing idealized standards for research in the engineering of instrumentation. Since instrumental technology lies at the core of laboratory studies of the micro-world, we look to the designs of such devices as expressions of standards for research. The laboratory techniques are laden with notions about an instrument's power, experimenter's skills, and specimen's properties. When explaining how valid signals are generated, enhanced, transformed, and converted into data, designers take a stance about knowledge acquisition, recommending how experiments should be performed through the conception of such technologies. A philosophy from design, not of design, emerges from a careful reading of design plans.

Designers deploy abstract ideals, theories, models, and physical laws as cognitive tools. The historical precedent for using ideas as cognitive instruments is quite clear. Mathematical machines were discovered and applauded for their computational power without actually producing a material product. For example, Gottfried Wilhelm Leibniz developed arithmetic machines without material features but with capacities to accomplish addition, subtraction, multiplication, division, and extraction of square roots. Such a machine served practical purposes but held no theoretical implications for arithmetic. In the 1840s, Charles Babbage wrote about the power of an analytical engine without actually resorting to a material device. In the twentieth century,

Alan Turing created a machine that generates a set of possible algorithms, responsive to a world of thought that is not limited to a world of silicon chips, copper wires, and electric circuits. As I argue in chapter 4, the engineering of laboratory instruments demonstrates how design plans resort to diagrammatic reasoning in the form of iconic modeling. Designers define an instrument through a vocabulary of pictorial symbols, through a shape-grammar of abstract concepts, and by leading readers through a thought experiment to demonstrate the ways in which the instruments function in research. The iconic modeling of engineering design demands skills and devices that are quite familiar to visual artists. Here again, vicarious witnessing is shown to have epistemic implications for the feasibility of design plans.

My proposals deviate from a familiar notion that any technology is reducible to its physicality. Dependency of an instrument's functionality on its physical structure is conveyed through engineering. In deploying a new laboratory technology, researchers need reassurance that a certain sequence of (desirable) analytical signals will be generated, transformed, and rendered accessible. Before researchers anticipate what artifacts will be produced in the laboratory, designers construct an idealized plan of an ordered sequence of changes. Predictability is essential, and fidelity to causal models of the physical sciences establishes predictability. A duality of function and structure emerges from the designers' conception of instrumentation. Underlying the connections among the experimental phenomena are causal capacities of materials to generate change. Technological interventions change the "natural" state of affairs by exploiting the capacities of certain preexisting forms. With scanning techniques of contemporary spectroscopy, chemists monitor the real, substantial changes that are technologically provoked. Such changes show the working of nature's nomological machines, as I argue in chapter 7. In current research, a specimen functions as a machine of nature, responsive to technological inducements, and operating reactively, not passively. A specimen's capacities are revealed through the new entities (states, events, or products) that are brought into existence during an experiment. Again, a pure specimen functions as a signal-producing machine, characterized by capacities for coding, transmitting, receiving, and decoding information.

Portions of the world that are revealed through such instruments exhibit affinities to devices of human construction. In the end, the machine/nonmachine duality is itself transformed by technological advances. Consider how the brain/machine distinction is blurred with the advent of the Turing machine and how nanostructures lie at the interface among solid-state physics, supramolecule chemistry, and molecular biology. The natural/nonnatural distinction is continually challenged by discoveries of new technologies. Design plans for instruments can be read for their philosophical stance, as a medium for ideas about inquiry in the pursuit of knowledge. They can also be read for the ontological categories of the accessible portions of the world and for the skills of experimenters that are enhanced through research. In this way, scientists are oriented to the detectable world through this integration of minds and tools in research.

NOTES

CHAPTER 1: SCIENCE, TECHNOLOGY, AND PHILOSOPHY

1. See Smith and Wise (1989) for an excellent study of the laboratory practices of Lord Kelvin.

CHAPTER 2: ANALOGIES OF DESIGN

1. Berzelius writes: "If we are seeking to form an idea about organic compounds, we have at the present time only a single method the sureness of which is beyond dispute. . . . I mean that we must take as our point of departure the comparison with inorganic compounds" (qtd. in Brooke 1980, p. 42).

2. Brooke provides an excellent study of the analogical reasoning associated with the development of organizing chemistry (1980).

3. The philosophical literature on scientific analogy is quite extensive (Black 1962, p. 239; Harré 1970, chap. 2; Hesse 1966; McMullin 1967; Spector 1965, pp. 135ff.; Swanson 1966, p. 306). Morrison and Morgan have recently examined how models themselves function as instruments of science (1999, chap. 2). In my opinion, the most compelling arguments for the centrality of analogical models in the empirical sciences appear in Harré's *Principles of Scientific Thinking* (1970). When an iconic model has a subject matter that is distinct from its source, the model is analogical.

4. As Niels Bohr has argued, a laboratory phenomenon is a novel physical event, revealing how an apparatus-world complex affords certain states under laboratory conditions (Harré 1986, p. 306).

5. In his Nobel lecture, C. T. R. Wilson provided the rationale for his cloud chamber (1965). This technique required the following steps: the supersaturation of air by sudden expansion of the gas, the passage of the ionizing particles through the supersaturated gas, and the illumination of the cloud condensed on the ions along the track.

6. In his search for numerical spectral relationships, Johan Jacob Balmer was impressed by Stoney's analogies to acoustics. Balmer suggested that the spectral lines of a given substance might be conceived as the overtones of the one "key-note" (McGucken 1969, p. 131). He believed that a simple law existed that would describe the hydrogen lines, which led to his discovery of the series formula.

7. I thank Eric Scerri for bringing this case to my attention.

8. The metaphoric underpinnings of chemistry are developed in wonderful detail

by T. L. Brown in *Making Truth: Metaphor in Science* (2003). I would like to thank one of the reviewers of my manuscript for bringing this excellent work to my attention.

CHAPTER 3: TESTING DESIGN PLANS

1. This has been a long-standing dream of science fiction writers, inspiring the 1957 movie *The Incredible Shrinking Man,* based on Richard Matheson's book of the same title. When the movie's lead character mysteriously shrinks, he is endangered by unfamiliar terrain and potential predators. To survive, he has to walk on, crawl on, or somehow maneuver around solid bodies as he quickly processes tactile clues from his surroundings.

2. The methodology includes the following components (Winsberg 1999, p. 288): a calculated structure for the theory, techniques of mathematical transformation, a choice of parameters, initial and boundary conditions, reduction of degrees of freedom, ad hoc models, a computer and a computer algorithm, a graphics system, and an interpretation of numerical and graphical output coupled with an assessment of their reliability.

CHAPTER 4: ICONS OF DESIGN AND IMAGES OF ART

1. Ludwig Wittgenstein becomes a valuable ally in a campaign to link pictorial language with visual thinking. He never abandons his fascination with pictures as a basis of thought. A sentence in a story gives us a picture in the mind's eye. His example of an engineer's drawings is quite telling: "What we call 'descriptions' are instruments for particular uses. Think of a machine-drawing, a cross-section, an elevation with measurements, which an engineer has before him" (1958, paragraph 291).

2. Through linguistic metaphor, what was impossible, inconceivable, and incoherent based on literal vocabulary becomes possible, conceivable, and coherent. This can be explained as a second-order interaction of conceptual fields, as I argue elsewhere (Rothbart 1997, chap. 2).

3. I am indebted to Alane Keller for bringing this to my attention.

4. This is evident in a recent analysis of Piet Mondrian's work: "[Mondrian] was fascinated by the invisible, especially as embodied in the concept of the fourth dimension. This he conceived as a space that lay beyond sensory perception, and that reversed our common-sense understanding of what was real and unreal. Such common-sense understanding, he believed, is based on the illusory three-dimensional universe, in which we exist on a lower level of consciousness" (Golding 2001, p. 12).

5. Egyptian artists often distorted the relative size of humans to depict visually the social status of adult males in comparison to adult females and children (Hagen 1986, p. 188).

CHAPTER 5: MICROSCOPES, MACHINES, AND MATTER

1. Another challenge to the medieval distinction between these two categories of instruments arises from studies of terrestrial magneticism at the close of the seventeenth century. Most investigators who studied terrestrial magnetism were mathematical practitioners, skilled in navigation and the production of instruments. The compass, for example, was hailed as one of the great wonders of the world. But magnetism also became an important subject for philosophical investigation, recognized as an abstract event of nature (Warner 1994, p. 72). In a further challenge to the distinction between mathematical and optical instruments, mathematical devices were equipped with telescopic sights, and optical instruments were equipped with substantial mounts and graduated circles. By the beginning of the nineteenth century, the term "philosophical instrument" was replaced by "scientific instrument" in France, Germany, and England (Warner 1990, pp. 85–88).

2. He writes: "In several of those vegetables, whilst green, I have with my microscope, plainly enough discovered those Cells or Pores filled with juices, and by degrees sweating them out" (Hooke 1961, p. 55).

3. In *The Posthumous Works of Robert Hooke*, he writes: "The Business of Philosophy is to find out a perfect Knowledge of the Nature and Properties of bodies, and of the Causes of natural productions" (Hooke qtd. in Waller 1705, p. 3).

4. Kant's philosophy of science has enjoyed a resurgence of interest in recent years, with special attention given to his *Critique of Judgment* (cf. Butts 1984 and 1993; Edwards 1991; Förster 1991; Friedman 1992; Kitcher 1995). In this work, Kant attempts to unify a system of experience with a system of nature. The attempt has inspired generations of scientists to examine the philosophical underpinnings of their disciplines. In the twentieth century, F. A. Paneth, an eminent experimental chemist and co-founder of radio chemistry, argues that the task of chemistry is to discover the transcendental world of basic substances (Paneth 1965, p. 152).

5. According to a familiar, but flawed, reading, Kant holds that actual scientific laws of nature are deductively inferred a priori from his transcendental method. On this interpretation, his transcendental principles have no currency to modern science, because they are overthrown by the empirical refutation of Newtonian physics. But this reading grossly distorts Kant's doctrines. Nothing in his transcendental philosophy implies the actual certitude of specific scientific laws, since he makes no attempt to deduce logically the laws of mechanics or any empirical law whatever (Buchdahl 1986, p. 156). His transcendental principles are a priori in the sense that they comprise preconditions for the possibility, but not the actuality, of scientific laws.

6. During the seventeenth and eighteenth centuries, natural philosophers wrote freely about the causal powers of material entities to produce empirical change. The sun has a causal power to warm a stone; the moon influences the tides; and arrangement of molecules causally determines the empirical attributes of a chemical

substance. According to Locke, primordial substance must be endowed with powers to produce ideas of secondary qualities of experience.

7. For an excellent study of Kant's developing insights toward the *Critique of Judgment,* see Scherer (1995).

8. I extend Kant's principle of purposiveness to certain philosophical aspects of contemporary instrumentation in Rothbart and Scherer (1997).

CHAPTER 6: ATOMS

1. Primas was long intrigued by the work of Norbert Wiener, Claude Shannon, Andrei Kolmogorov, A. I. Khintchine, and others on stochastic processes in the context of electronic measurement and data transmission. The work of these major figures in stochastic system theory, information theory, and cybernetics greatly advanced the state of knowledge about ways in which signals can be filtered from noise. Primas relied heavily on optimized data processing and filtering theory to better understand the character of weak and noisy signals in NMR spectrometers. Random noise had been used to characterize technological artifacts. Primas's original idea was that the same apparatus can be used in testing physical quantum mechanical systems. Primas provided, for the first time, an account of the response of a nuclear spin system to a stochastic perturbation. Through the creative use of knowledge from separate domains, he relied on linear systems response theory, Fourier and Laplace transformation theory, and the associated frequency domain methods for determining system response and stability.

2. I thank John Schreifels for his valuable insight concerning the history of scanning techniques of surface chemistry, leading to the information presented in this section. His analysis appears separately in a co-authored work (Rothbart and Schreifels 2005).

3. For a history of this invention, see Drechsler (1978).

CHAPTER 7: SPECIMENS AS MACHINES

1. During the time of the ancients, the machine metaphor was evident in creation myths. According to Greek myth, the capricious behavior of the gods was responsible for cosmic events. The occurrence of good and evil in the world exhibited the moral character of the gods, which drew upon analogies to human virtues and vices. Of course, under some circumstances, such as ritualistic sacrificing, the gods could be influenced. But in general, cosmic forces were seen as untamed, uncontrollable, and unpredictable. Ancient experiments were performed only on the segments of nature that were tamed, controlled, and predictable. To carry out an experiment, human agents would tamper with the natural order through acts of agitation, manipulation, or intervention. Ancient Greeks of the fifth century B.C. analogized their idea of causal responsibility to the human capacity to produce an injustice. An agent disturbs a state

of equilibrium and is thus responsible for some evil in nature. Similarly, in terms of causal responsibility, an agent initiates a sequence of events and by performing an action becomes responsible for consequences for good or ill.

2. Keller provides an excellent analysis of the study of cellular automata as simulations of global, or macro-dynamics, properties (2002, chap. 9).

3. In this context, Salmon identifies causation as a continuous process analogous to a filament or thread but not to a chain. For Salmon, a causal process is one that transmits a "mark" so that certain characteristics are modified as a result of the process. A mark is a modification, understood as a transfer of structure. A material particle can be marked by a change of its characteristics (1998, p. 250). The transmission of light from one place to another is a causal process. Any kind of electromagnetic radiation is causal, since radiation has a capacity to leave a mark (p. 16). An electromagnetic signal transmits a causal influence from a control device to a mechanism that raises a door, for example, and an interaction occurs between the signal and that mechanism. Dowe (1992) defines a causal process as preservation of conserved quantities. For example, when an atom absorbs a photon, energy is transferred as a conserved quantity, based on conservation theories of physical science. For Dowe, causality can be understood as the transference of a conserved quantity along a world line. A causal process is, by definition, a world line of an object that manifests a conserved quantity (p. 210).

4. Keller provides an illuminating analysis of Thompson's contributions and the unpleasant reception of his work by biologists (2002, chap. 2).

5. The variations of twist, roll, and tilt between one base pair and its successor have been summarized in findings of fluctuations of a particular strand of DNA (Sarai et al. 1989, p. 7844). The fluctuation of helical twist is lowest for the adjacent base pairs involving the guanine-guanine sequence, with a range of ±10.7 degrees. Helical twist is highest for the thymine-adenine adjacent base pairs, ranging from -24 degrees to +24 degrees. The fluctuations of roll show only a 5.2 degree fluctuation for the guanine-guanine adjacent base pairs, but the fluctuation of roll is relatively great, at 10.7 degrees, for the thymine-adenine sequence. Tilt is less flexible than twist and roll. The thymine-adenine base pair and successor has the lowest degree of fluctuation in roll of 3.3 degrees, while the adenine-adenine sequence has the highest degree of fluctuation in roll at 5.1 degrees (pp. 7844–45).

6. Local flexibility of the double helix is conceived through dimer sequence energies between atoms of one residue I and those of an adjacent residue $I+1$. These energies are a sum of pairwise van der Waals repulsions, London attractions, and electrostatic interactions between the jth atoms of residue I and the kth atoms of residue $I+1$. In this respect, the molecule's twist, roll, and tilt are conceived biophysically, as the dimer sequence energies. Nonbonded parameters include London constants, which are evaluated from the atomic polarizabilities and effective numbers of valence electrons of atoms j and k (Srinivasan et al. 1987, pp. 461–63).

7. I would like to thank Michael Akeroyd for bringing this to my attention.

GLOSSARY

ABSORPTION SPECTROMETER is an instrument of analytical chemistry used for identification, structure elucidation, and quantification of organic compounds. Depending on the molecular structure of the sample, various wavelengths of radiation are absorbed, reflected, or transmitted. That part of the radiation that passes through the sample is detected and converted to an electrical signal, constituting an event of the phenomenal realm.

AESTHETIC MOVEMENT is an aesthetic experience in which sensory perceptions entice a viewer to anticipate possible images that can be projected onto a scene.

ANALYTICAL INSTRUMENT, commonly used for chemical analysis, converts an analytical signal that is usually not directly detectable to a form that is detectable.

ANTHROPOMORPHISM is attribution of a human form to that which is not human, such as the Deity.

AIR PUMP is a machine for exhausting air out of a pump by means of the movement of a piston, also known as a pneumatic engine.

BEER'S LAW holds that the absorbance of a species in a solution is directly proportional to the path length through the solution and the concentration of the absorbing species.

CAD is an automated (computer-based) method for generating graphics for designing and drafting.

CAUSAL MECHANISM is, according to seventeenth-century natural philosophers, a system of entities and processes with capacities to produce changes in other things under suitable conditions, or to have changes produced in them by other things. For some of the mechanistic philosophers of the seventeenth century, the smallest mechanisms are tiny wheels and springs of nature.

CHROMATIC ABERRATION is a defect of microscopes in which the image is surrounded by colored edges. This defect is caused by the refraction of different wavelengths of white light.

COMPOUND MICROSCOPE is a microscope that includes a combination of two or more lenses, which are usually convex.

CONFORMATIONAL FLEXIBILITY OF DNA refers to possible movements of one base pair in relation to another base pair, with respect to the properties of helical twist, tilt, roll, and propeller twist.

DEPTH OCCLUSION is a process of hiding or exposing surfaces as a result of changing one's scale of measurement.

DESIGN SPACE is an abstract representational space used to replicate movement, where certain conceptual elements are imagined and the effects are anticipated.

DESIRABLE SIGNAL is an information-carrying variable that has its source in a specimen's properties.

DIAGRAMMATIC REASONING is, according to C. S. Peirce, a visualizable mode of problem solving, which requires the introduction of new elements to define a problem, the use of elements as general concepts, and the creation of hypotheses for testing.

ENERGY BARRIER is a quantity of energy equal to the difference between an electron attached to the surface atom and an electron that is freed from the influence of a surface atom.

EVOLUTIONARY THEORY OF PERCEPTION, as advocated by John Dewey, Karl Popper, and J. J. Gibson, holds that sensory experiences occur when a perceiver is actively planning and engaging in processes of the environment.

EXTRAMISSION THEORY OF OPTICS, advanced by Euclid and other ancients, holds that vision occurs when radiation is emitted from the eye and sent to "feel" the properties of visible objects.

FIELD ION MICROSCOPY is an instrumental technique in which the atoms of a specimen adsorb helium or neon atoms, followed by the application of a potential on the order of 10,000 volts to the surface. The distances between neighboring ions released from this process are then measured.

GRAPHIC OCCLUSION is occlusion of edges or surfaces of bodies in a design space.

KINESTHETISM is an ability to acquire experiences of an object's properties concurrently through visual, verbal, or mathematical techniques.

INTRAMISSION THEORY OF OPTICS holds that vision occurs when radiation is emitted from a visual body and transmitted through a medium, then impinges on the surface of the eye.

MATERIAL SKILL is capacity of human agents to produce a desirable product or state based on a manipulation of material bodies, an assessment of risks and benefits of certain action, and an evaluation of the results.

MATHEMATICAL INSTRUMENT is a device of the ancients and moderns used in practical affairs for weighing, measuring, and otherwise attaching numbers to properties of materials.

MECHANISTIC PHILOSOPHY of the seventeenth century holds that sensible phenomena are effects of purely mechanical interactions of particles, interactions that are in principle characterized by mathematical properties.

NATURAL MAGIC is the practice of the seventeenth and eighteenth centuries of producing startling sensory phenomena through tricks that disguise causal production of the phenomena.

NOISY SIGNAL is an information-carrying variable that has an "extraneous" source in relation to a specimen's properties, leading to contamination of the resulting data.

NOMOLOGICAL MACHINE is, according to Nancy Cartwright, any material body that exhibits a kind of regular behavior due to the lawlike capacities of a causal mechanism.

OCCLUDING EDGE is an edge that is adjacent to a surface that can go out of sight or return into sight, depending upon a change in an observer's relative position.

OCCLUDING SURFACE is a surface that is adjacent to a volume that can go out of sight or return into sight, depending upon a change in an observer's relative position.

ORTHOGRAPHIC PROJECTION is a projection system in which optical rays from the surfaces of a material body to a picture plane are parallel to each other and intersect the picture plan at right angles.

PHILOSOPHICAL INSTRUMENT, in the context of seventeenth-century experiments, is a device designed to reveal the true laws of nature, disclosing otherwise hidden processes that were causally responsible for empirical phenomena.

POWER OF AN AGENT is disposition of an agent to act in ways that produce an effect on another person or material thing.

PRINCIPLE OF REVERSIBLE OCCLUSION is James J. Gibson's thesis that for any movement of a point of observation that hides previously exposed surfaces, there is an opposite movement that reveals them.

PRINCIPLE OF CONGRUITY is Robert Hooke's thesis that every material body is endowed with a capacity to unite with some other body, depending on the physical attributes of both bodies.

RASTER is a scanning pattern associated with scanning electron microscopy in which an electron beam sweeps across a surface in a straight line, returns to its starting position, and shifts downward by a standard increment.

REDUCTIVE PHYSICALISM holds that all detectable phenomena are explainable through the dynamics of energy exchange, according to the laws of physics.

SPHERICAL ABERRATION is a defect of microscopes in which a slight blurring effect occurs due to the spherical curvature of the lens.

SCANNING ELECTRON MICROSCOPY is an instrumental technique in which the surface atoms of a metal are bombarded with high-energy electrons, followed by measurement of the resulting spatial distribution of electrons.

SCANNING TUNNELING MICROSCOPY is an instrumental technique in which a needle rides across a prepared surface, passing a current from the probe to the surface of the material under examination, creating a "tunneling" effect in which electrons escape the influence of atoms by traveling on a different route from its "normal" path without experimental interferences.

SHAPE-GRAMMAR is a set of rules found typically in engineering and architecture for conveying information in a design plan through the proper use of points, lines, and shapes.

SPECTROSCOPY is the study of instrumental methods used for revealing properties of a sample, based on a physical interaction between radiation and a sample's molecules, atoms, or nuclei.

THOUGHT EXPERIMENT is a "mental manipulation" of apparatus, specimen, and laboratory conditions for the purpose of anticipating the results of a material test, whether or not the test is actually performed.

TOOL, in the material sense, is a detached material object that can be controlled by a user to perform work in the mechanical sense of transferring energy.

TRANSCENDENTAL PRINCIPLE, according to Immanuel Kant, conveys the preconditions that must be met for the empirical laws in the proper sciences, such a Newtonian mechanics, to be valid.

TUNNELING EFFECT is a physical process in which electrons are freed from a potential barrier without reaching the level of thermal energy that is normally required. Designers of the scanning tunneling microscope exploited the process.

VIRTUAL REALITY MODELING LANGUAGE is a language enabling designers to describe three-dimensional bodies in an object-oriented manner.

VIRTUAL WITNESSING is a process in which a reader of experimental reports vicariously reenacts significant features of an experiment, focusing on an instrument's design, material apparatus, and microscopic phenomena. According to S. Shapin and S. Schaffer, many of Robert Boyle's experiments were validated through virtual witnessing.

REFERENCES

Ackermann, R. 1985. *Data, Instruments and Theory.* Princeton, N.J.: Princeton University Press.

Agazzi, E. 1999. "From Technique to Technology: The Role of Modern Science." *Techné* 4, no. 2.

Arnheim, R. 1964. *Art and Visual Perception: A Psychology of the Creative Eye.* Berkeley: University of California Press.

Atkins, P. W. 1978. *Physical Chemistry.* San Francisco: W. H. Freeman.

Baigrie, B. 1996. "Descartes's Scientific Illustrations and 'la grande mécanique de la nature.'" In *Picturing Knowledge: Historical and Philosophical Problems Concerning the Use of Art in Science,* ed. B. Baigrie, 86–134. Toronto: University of Toronto Press.

Baird, D. 1991. "Baird Associates' Commercial Three-Meter Grating spectrograph and the Transformation of Analytical Chemistry." *Rittenhouse* 5, no. 3:65–80.

———. 1998. "Scientific Instrument Making, Epistemology, and the Conflict between Gift and Commodity Economies." *Techné* 4, no. 2.

———. 2000a. "Analytical Instrumentation and Instrumental Objectivity." In *Of Minds and Molecules: New Philosophical Perspectives on Chemistry,* ed. N. Bhushan and S. Rosenfeld, 90–113. Oxford: Oxford University Press.

———. 2000b. "Encapsulating Knowledge." *Foundations of Chemistry* 2, no. 1:6–46.

———. 2002. "Histories of Baird Associates." In *From Classical to Modern Chemistry: The Instrumental Revolution,* ed. P. J. T. Morris, 129–48. London: Royal Society of Chemistry.

———. 2004. *Thing Knowledge: A Philosophy of Scientific Instruments.* Berkeley: University of California Press.

Baird, D., and T. Faust. 1990. "Scientific Instrument, Scientific Progress and the Cyclotron." *British Journal for the Philosophy of Science* 41:147–76.

Baird, D., and A. Nordmann. 1994. "Facts-Well-Put." *British Journal for the Philosophy of Science* 45:37–77.

Baker, G. P., and P. M. S. Hacker. 1980. *Wittgenstein: Understanding and Meaning.* Chicago: University of Chicago Press.

———. 1985. *Wittgenstein: Rules, Grammar and Necessity.* Oxford: Basil Blackwell.

Barnes, B., and B. Bloor. 1982. "Relativism, Rationalism and the Sociology of Knowledge." In *Rationality and Relativism,* ed. M. Hollis and S. Lukes, 21–47. Cambridge, Mass.: MIT Press.

Barnes, B., and J. Henry. 1996. *Scientific Knowledge: A Sociological Analysis.* Chicago: University of Chicago Press.

Baroni, Constantino, ed. 1956. *Leonardo da Vinci.* New York: Reynal.

Bechtel, W., and R. Richardson. 1993. *Discovering Complexity: Decomposition and Localization as Strategies in Scientific Research.* Princeton, N.J.: Princeton University Press.

Bennett, J. A. 1986. "The Mechanics' Philosophy and the Mechanical Philosophy." *History of Science* 24:1–28.

Bertolline, G. 1988. "The Role of Computers in the Design Process." *Engineering Design Graphics Journal* 52, no. 2:18–22, 30.

Binnig, G., and H. Röhrer. 1982. Scanning Tunneling Microscope. U.S. Patent 4,343,993, filed August 10, 1982. Assignee: International Business Machines Corporation, Armonk, N.Y.

———. 1985. "The Scanning Tunneling Microscope." *Scientific American* 182 (August): 50–56.

Binnig, G., et al. 1982. "Tunneling through a Controllable Vacuum Gap." *Applied Physics Letters* 40, no. 2:178–80.

Black, M. 1962. *Models and Metaphors.* Ithaca, N.Y.: Cornell University Press.

Bogen, J., and J. Woodward. 1988. "Saving the Phenomena." *Philosophical Review* 97:303–52.

Bonnel, D., ed. 1993. *Scanning Tunneling Microscopy and Spectroscopy: Theory, Techniques and Applications.* New York: VCH Publishers.

Bottomley, L. A., J. E. Coury, and P. N. First. 1996. "Scanning Probe Microscopy." *Analytical Chemistry* 68:185R-230R.

Bradbury, S. 1967. *The Evolution of the Microscope.* Oxford: Pergamon Press.

Brandon, R. 1985. "Grene on Mechanism and Reductionism: More Than Just a Side Issue." In *PSA 1984,* vol. 2, ed. P. Asquith and P. Kitcher, 345–53. East Lansing, Mich.: Philosophy of Science Association.

Brock, W. H. 1992. *The Norton History of Chemistry.* New York: W. W. Norton.

Brooke, J. 1973. "Chlorine Substitution and the Future of Organic Chemistry." *Studies in the History and Philosophy of Science* 4:47–94.

———. 1980. "The Chemistry of the Organic and the Inorganic." *Kagakushi: Journal of the Japanese Society of the History of Chemistry* 7:37–60.

Brown, H. 1985. "Galileo on the Telescope and the Eye." *Journal of the History of Ideas* 46:487–501.

Brown, J. R. 1991. *The Laboratory of the Mind.* New York: Routledge.

Brown, T. L. 2003. *Making Truth: Metaphor in Science.* Urbana: University of Illinois Press.

Buchdahl, G. 1986. "Kant's 'Special Metaphysics' and *The Metaphysical Foundations of Natural Science.*" In *Kant's Philosophy of Physical Science,* ed. R. Butts, 127–61. Dordrecht-Holland: D. Reidel.

Burian, R. 1996. "Underappreciated Pathways toward Molecular Genetics as Illustrated by Jean Brachet's Cytochemical Embryology." In *The Philosophy and History of Molecular Biology: New Perspectives,* ed. S. Sarkar, 671–85. Dordrecht-Holland: Kluwer Academic Publishers.

Burtt, E. A. 1992. *The Metaphysical Foundations of Modern Science.* Atlantic Highlands, N.J.: Humanities Press.

Butts, R. E. 1984. *Kant and the Double Government Methodology: Supersensibility and Method in Kant's Philosophy of Science.* Dordrecht-Holland: D. Reidel.

———. 1993. *Historical Pragmatics: Philosophical Essays.* Dordrecht-Holland: Kluwer Academic Publishers.

Capek, M. 1961. *The Philosophical Impact of Contemporary Physics.* New York: Van Nostrand Reinhold.

Carroll, Lewis. 1964. *Alice's Adventures in Wonderland: Through the Looking-glass.* Garden City, N.Y.: International Collectors Library. (Orig. pub. 1896.)

Cartwright, N. 1999. *The Dappled World: A Study of the Boundaries of Science.* Cambridge: Cambridge University Press.

Cassirer, E. 1923. *"Substance and Function" and "Einstein's Theory of Relativity."* New York: Dover.

Chapman, Jan. 1999. *The Art of Rhinoceros Horn Carving in China.* London: Christie's Books.

Charleston, R. J., and L. M. Angus-Butterworth. 1957. "Glass." In *A History of Technology: Volume III, From the Renaissance to the Industrial Revolution c1500–c1750,* ed. C. Singer, E. J. Holmyard, A. R. Hall, and T. I. Williams, 206–44. New York: Oxford University Press.

Chen, C. J. 1993. *Introduction to Scanning Tunneling Microscopy.* Oxford Series in Optical and Imaging Sciences. New York: Oxford University Press.

Coates, J. 1997. "Vibrational Spectroscopy: Instrumentation for Infrared and Raman Spectroscopy." In *Analytical Instrumentation Handbook,* ed. G. Ewing, 393–555. 2nd ed. New York: Marcel Dekker.

Collins, H. M., and S. Yearley. 1992. "Epistemological Chicken." In *Science as Practice and Culture,* ed. A. Pickering, 301–26. Chicago: University of Chicago Press.

Coor, T. 1968. "Signal to Noise Optimization in Chemistry—Part One." *Journal of Chemical Education* 45, no. 7 (July): A533.

Currell, G. 1987. *Instrumentation.* London: Wiley.

Del Re, G. 2000. "Models and Analogies in Science." *Hyle: An International Journal for the Philosophy of Chemistry* 6:5–15.

Descartes, René. 2001. *Discourse in Methods, Optics, Geometry, and Meteorology.* Trans. P. J. Olscamp. Indianapolis: Hackett. (Orig. pub. 1637.)

Dewey, J. 1958. *Experience and Nature.* New York: Dover.

Dickerson, R., H. Drew, and B. Conner. 1981. "Single-Crystal X-Ray Structure Analyses of A, B, and Z Helices or One Good Turn Deserved Another." In *Biomolecular Stereodynamics,* vol. 1, ed. R. Sarma, 1–34. New York: Adenine Press.

Dietz, T., and T. R. Burns. 1992. "Human Agency and the Evolutionary Dynamics of Culture." *Acta Sociologica* 35:187–200.

Dorst, K., and J. Dijkhuis. 1995. "Comparing Paradigms for Describing Design Activity." *Design Studies* 16:261–74.

Dowe, P. 1992. "Wesley Salmon's Process Theory of Causality and the Conserved Quantity Theory." *Philosophy of Science* 59:195–216.

Drechsler, M. 1978. "Erwin Müller and the Early Development of Field Emission Microscopy." *Surface Science* 70:1–18.

Dretske, F. 1981. *Knowledge and the Flow of Information.* Cambridge, Mass.: MIT Press.

Dubery F., and J. Willats. 1983. *Perspective and Other Drawing Systems.* New York: Van Nostrand Reinhold.

Duncan, H. 1986. "Kant's Methodology: Progress beyond Newton?" In *Kant's Philosophy of Physical Science,* ed. R. Butts, 273–306. Dordrecht-Holland: D. Reidel.

Earle, J. 1994. *Engineering Design Graphics.* 8th ed. Reading, Mass.: Addison-Wesley.

Edwards, B. J. 1991. "Der Ätherbeweis des Opus postumum und Kants 3. Analogie der Erfahrung." In *Übergang, Untersuchungen zum Spätwerk Immanuel Kants,* ed. Forum für Philosophie Bad Homburg, 77–104. Frankfurt: Vittorio Klostermann.

Eliade, M. 1954. *The Myth of the Eternal Return or, Cosmos and History.* Princeton, N.J.: Princeton University Press.

Emirbayer, M., and A. Mische. 1998. "What Is Agency?" *American Journal of Sociology* 103, no. 4:962–1023.

Emmerik, J. G. M. van. 1991. "A Direct Manipulation Environment for Conceptual Design of Three-Dimensional Objects." In *Advances in Object-Oriented Graphics I,* ed. E. H. Blake and P. Wisskirchen, 149–74. Berlin: Springer-Verlag.

Fair, D. 1979. "Causation and the Flow of Energy." *Erkenntnis* 14:219–50.

Ferguson, E. 1992. *Engineering and the Mind's Eye.* Cambridge, Mass.: MIT Press.

Fernández, E. 1993. "From Peirce to Bohr: Theorematic Reasoning and Idealization in Physics." In *Charles S. Peirce and the Philosophy of Science,* ed. E. Moore, 233–45. Tuscaloosa: University of Alabama Press.

Förster, E. 1991. "Idee des Übergangs, Überlegungen zum Elementarsystem der bewegenden Kräfte." In *Übergang, Untersuchungen zum Spätwerk Immanuel Kants,* ed. Forum für Philosophie Bad Homburg, 28–48. Frankfurt: Vittorio Klostermann.

Franklin, A. 1986. *The Neglect of Experiment.* Cambridge: Cambridge University Press.

Frederick, C., J. Grable, M. Melia, C. Samudzi, L. Jen-Jacobson, B. Wang, P. Greene, H. Boyer, and J. Rosenberg. 1984. "Kinked DNA in Crystalline Complex with EcoRI Endonuclease." *Nature* 309 (May): 327–30.

Friedman, M. 1992. *Kant and the Exact Sciences.* Cambridge, Mass.: Harvard University Press.

Galison, P. 1987. *How Experiments End.* Chicago: University of Chicago Press.

————. 1997. *Image and Logic: A Material Culture of Microphysics.* Chicago: University of Chicago Press.

Galison, P., and A. Assmus. 1989. "Artificial Clouds and Real Particles." In *The Uses of Experiment: Studies in the Natural Sciences,* ed. D. Gooding, T. Pinch, and S. Schaffer, 225–74. Cambridge: Cambridge University Press.

Genova, J. 1995. *Wittgenstein: A Way of Seeing.* New York: Routledge.

Gentner, D. 1982. "Structure-Mapping: A Theoretical Framework for Analogy." *Cognitive Science* 7:155–70.

Gibson, J. J. 1986. *The Ecological Approach to Visual Perception.* Hillsdale, N.J.: Lawrence Erlbaum.

Golding, J. 2001. "Paths to the Absolute: Malevich, Kandinsky, Pollock, Newman, Rothko, and Still." *New York Review of Books.* April 26, p. 12.

Goldschmidt, G. 1997. "Capturing Indeterminism: Representation in the Design Problem Space." *Design Studies* 18, no. 4:441–55.

Gooding, D. 1989. "'Magnetic Curves' and the Magnetic Field: Experimentation and Representation in the History of a Theory." In *The Uses of Experiment: Studies in the Natural Sciences,* ed. D. Gooding, T. Pinch, and S. Schaffer, 183–224. Cambridge: Cambridge University Press.

————. 1990. *Experiment and the Making of Meaning.* Dordrecht: Kluwer Academic Publishers.

————. 1992. "Putting Agency Back into Experiment." In *Science as Practice and Culture,* ed. A. Pickering, 65–112. Chicago: University of Chicago Press.

Gouk, P. M. 1980. "The Role of Acoustics and Music Theory in the Scientific Work of Robert Hooke." *Annals of Science* 37:559–85.

Hacker, P. M. S. 1986. *Insight and Illusion: Themes in the Philosophy of Wittgenstein.* Rev. ed. Oxford: Clarendon Press.

Hacking, I. 1983. *Representing and Intervening.* Cambridge: Cambridge University Press.

————. 1992a. "Disunified Sciences." In *The End of Science? Attack and Defense,* ed. R. Q. Elvee, 33–51. Lanham, Md.: University Press of America.

————. 1992b. "The Self-Vindication of the Laboratory Sciences." In *Science as Practice and Culture,* ed. A. Pickering, 29–64. Chicago: University of Chicago Press.

Hagen, M. 1986. *Varieties of Realism: Geometries of Representational Art.* Cambridge: Cambridge University Press.

Hankins, T. L., and R. J. Silverman. 1995. *Instruments and the Imagination.* Princeton, N.J.: Princeton University Press.

Harré, R. 1970. *The Principles of Scientific Thinking.* Chicago: University of Chicago Press.

————. 1984. *Personal Being.* Cambridge, Mass.: Harvard University Press.

————. 1985. "Creativity in Science." In *The Concept of Creativity in Science and Art,* ed. D. Dutton and M. Kraus, 19–46. Dordrecht: Martinus Nijhoff.

————. 1986. *Varieties of Realism.* Oxford: Basil Blackwell.

————. 1998. "Recovering the Experiment." *Philosophy* 73:353–77.

Harré, R., and E. H. Madden. 1977. *Causal Powers.* Oxford: Basil Blackwell.

Harwood, J. T. 1989. "Rhetoric and Graphics in *Micrographia.*" In *Robert Hooke: New Studies,* ed. M. Hunter and S. Schaffer, 119–47. Wolfeboro, N.H.: Boydell Press.

Henderson, K. 1999. *On Line and On Paper: Visual Representations, Visual Culture, and Computer Graphics in Design Engineering.* Cambridge, Mass.: MIT Press.

Hesse, M. 1966. *Models and Analogies in Science.* Notre Dame, Ind.: University of Notre Dame Press.

Hooke, R. 1961. *Micrographia or Some Physiological Descriptions of Minute Bodies Made by Magnifying Glasses with Observations and Inquiries Thereupon.* New York: Dover. (Orig. pub. 1665.)

Ingold, T. 1986. *Evolution and Social Life.* Cambridge: Cambridge University Press.

————. 2000. *The Perception of the Environment: Essays on Livelihood, Dwelling and Skill.* London: Routledge.

Johnson, A. 2001a. "Design and the Engineering Community: How Knowledge Was Produced in the Case of Antilock Braking Systems." Unpublished manuscript.

————. 2001b. "Virtual Tools: The Epistemological and Social Issues of Computer-Aided Chemical Process Design." Unpublished manuscript.

Johnson, A. L. 1990. "Functional Modelling: A New Development in Computer-Aided Design." In *Intelligent CAD II: Proceedings of the IFIP TC 5/WG 5.2 Second Workshop on Intelligent CAD,* ed. H. Yoshikawa and T. Holden, 203–12. North-Holland: Elsevier Science Publishers.

Johnson, M. 1987. *The Body in the Mind.* Chicago: University of Chicago Press.

Kandinsky, W. 1979. *Point and Line to Plane.* New York: Dover.

Kant, I. 1965. *Critique of Pure Reason.* Trans. Norman Kemp Smith. New York: St. Martin's Press. (Orig. pub. 1787.)

————. 1985. *Metaphysical Foundations of Natural Science.* Trans. J. W. Ellington. Indianapolis: Hackett. (Orig. pub. 1786.)

————. 1987. *Critique of Judgment.* Trans. W. Pluhar. Indianapolis: Hackett. (Orig. pub. 1790.)

————. 1993. *Opus Postumum.* Trans. E. Förster. Cambridge: Cambridge University Press.

Kapoor, S. 1969. "The Origins of Laurent's Organic Classification." *Isis* 60:477–527.

Keller, E. F. 2002. *Making Sense of Life: Explaining Biological Development with Models, Metaphors, and Machines.* Cambridge, Mass.: Harvard University Press.

————. 2003. "Models, Simulations, and 'Computer Experiments.'" In *The Philosophy of Scientific Experimentation,* ed. H. Radder, 198–215. Pittsburgh: University of Pittsburgh Press.

Kemp, M. 1996. "Temples of the Body and Temples of the Cosmos: Vision and Visualization in the Vesalian and Copernican Revolutions." In *Picturing Knowledge:*

Historical and Philosophical Problems Concerning the Use of Art in Science, ed. B. Baigrie, 40–85. Toronto: University of Toronto Press.

Kitcher, P. 1995. "Revisiting Kant's Epistemology: Skepticism, Apriority and Psychologism." *Nous* 29:285–315.

Knorr-Cetina, K. 1992. "The Couch, the Cathedral, and the Laboratory: On the Relationship between Experiment and Laboratory in Science." In *Science as Practice and Culture,* ed. A. Pickering, 113–38. Chicago: University of Chicago Press.

Kroes, P. 2003. "Physics, Experiments, and the Concept of Nature." In *The Philosophy of Scientific Experimentation,* ed. H. Radder, 68–86. Pittsburgh: University of Pittsburgh Press.

Kutschmann, W. 1986. "Scientific Instruments and the Senses: Towards an Anthropological Historiography of the Natural Sciences." *International Studies in the Philosophy of Science* 1:106–23.

Langer, S. 1942. *Philosophy in a New Key: Study in the Symbolism of Reason, Rite, and Art.* Cambridge, Mass.: Harvard University Press.

Laszlo, P. 1998. "Chemical Analysis as Dematerialization." *Hyle: An International Journal for the Philosophy of Chemistry* 4:29–38.

Latour, B. 1987. *Science in Action.* Cambridge, Mass.: Harvard University Press.

Layton, E. 1991. "A Historical Definition of Engineering." In *Critical Perspectives on Nonacademic Science and Engineering,* ed. P. Durbin, 60–79. Research in Technology Studies, vol. 4. Bethlehem, Penn.: Lehigh University Press.

Lindberg, D. C. 1992. *The Beginnings of Western Science.* Chicago: University of Chicago Press.

Machamer, P., L. Darden, and C. F. Craver. 2000. "Thinking about Mechanisms." *Philosophy of Science* 67 (March): 1–25.

Mahner, M., and M. Bunge. 1997. *Foundations of Biophilosophy.* Berlin: Springer-Verlag.

Mainzer, K. 1997. "Symmetry and Complexity." *Hyle: An International Journal for the Philosophy of Chemistry* 3:29–49.

———. 1999. "Computational Models and 'Virtual Reality.'" *Hyle: An International Journal for the Philosophy of Chemistry* 5:135–44.

Makkreel, R. 1990. *Imagination and Interpretation in Kant.* Chicago: University of Chicago Press.

Malmstadt, H. W., and C. G. Enke. 1963. *Electronics for Scientists.* New York: W. A. Benjamin.

Mann, C., T. Vickers, and W. Gulick. 1974. *Basic Concepts in Electronic Instrumentation.* New York: Harper and Row.

Maroun, R. and W. Olson. 1988. "DNA. II. Configurational Statistics of Rodlike Chains." *Biopolymer* 27:561–84.

Mathews, C. K., and K. E. van Holde. 1996. *Biochemistry.* Menlo Park, Calif.: Benjamin/Cummings.

McDonnell, J. 1997. "Descriptive Models for Interpreting Design." *Design Studies* 18, no. 4:457–73.

McGucken, W. 1969. *Nineteenth-Century Spectroscopy.* Baltimore: Johns Hopkins University Press.

McLaughlin, P. 2001. "Toward an Ecology of Social Action: Merging the Ecological and Constructivist Traditions." Unpublished manuscript.

McMullin, E. 1967. "What Do Physical Models Tell Us?" In *Logic, Methodology and Philosophy of Science III: Proceedings of the Third International Congress for Logic, Methodology and Philosophy of Science,* ed. B. Van Rootselaar and J. Staal, 384–96. Amsterdam: North-Holland.

Merleau-Ponty, M. 1993. "Eye and Mind." In *The Merleau-Ponty Aesthetics Reader: Philosophy and Painting,* ed. G. A. Johnson and M. B. Smith, 121–58. Evanston, Ill.: Northwestern University Press.

Miller, G., and P. Johnson-Laird. 1976. *Language and Perception.* Cambridge, Mass.: Harvard University Press.

Mitcham, C. 1994. *Thinking through Technology: The Path between Engineering and Philosophy.* Chicago: Chicago University Press.

Morris, P. J. T., ed. 2002. *From Classical to Modern Chemistry: The Instrumental Revolution.* London: Royal Society of Chemistry.

Morris, P. J. T., and A. Travis. 2002. "The Role of Physical Instrumentation in Structural Organic Chemistry in the Twentieth Century." In *From Classical to Modern Chemistry: The Instrumental Revolution,* ed. P. J. T. Morris, 57–86. London: Royal Society of Chemistry.

Morrison, M., and M. Morgan. 1999. *Models as Mediators: Perspectives in Natural and Social Science.* Cambridge: Cambridge University Press.

Müller, E. W., and T. T. Tsong. 1969. *Field Evaporation: Field Ion Microscopy.* New York: American Elsevier.

Newton, I. 1952. *Opticks.* Based on 4th ed., 1730. New York: Dover.

———. 1959. "Newton to Oldenburg: 7 December 1675." In *The Correspondence of Isaac Newton, Volume I: 1665–1675,* ed. H. Turnbull, 362–89. Cambridge: Cambridge University Press.

Olson, W., N. Marky, R. Jernigan, and V. Zhurkin. 1993. "Influence of Fluctuations on DNA Curvature: A Comparison of Flexible and Static Wedge Models of Intrinsically Bent DNA." *Journal of Molecular Biology* 232:530–54.

Oxman, R. 1997. "Design by Re-representation: A Model of Visual Reasoning in Design." *Design Studies* 18, no. 4:329–47.

Pahl, G., and W. Beitz. 1988. *Engineering Design: A Systematic Approach.* Ed. K. Wallace. Berlin: Springer-Verlag.

Paneth, F. A. 1962. "The Epistemological Status of the Chemical Conception of Element." *British Journal for the Philosophy of Science* 13 (May): 1–14, 144–60.

———. 1965. "Chemical Elements and Primordial Matter: Mendeleeff's View and the Present Position." In *Chemistry and Beyond,* ed. H. Dingle and G. R. Martin, 53–72. New York: Wiley.

Parsons, M. 1997. "Atomic Absorption and Flame Emission Spectrometry." In *Analytical Instrumentation Handbook,* ed. G. W. Ewing, 2nd ed., 257–326. New York: Marcel Dekker.

Passini, R. 1996. "Wayfinding Design: Logic, Application and Some Thoughts on Universality." *Design Studies* 17:319–31.

Peirce, C. S. 1976. *The New Elements of Mathematics.* Ed. C. Eisele. 4 vols. The Hague: Mouton.

Petersen, A. 2000. "Philosophy of Climate Science." *Bulletin of the American Meteorological Society* 81, no. 2:265–71.

Pickering, A. 1989. "Living in the Material World: On Realism and Experimental Practice." In *The Uses of Experiment: Studies in the Natural Sciences,* ed. D. Gooding, T. Pinch, and S. Schaffer, 275–97. Cambridge: Cambridge University Press.

———. 1995. *The Mangle of Practice: Time, Agency, and Science.* Chicago: University of Chicago Press.

Plaass, P. 1965. *Kant's Theory of Natural Science.* Trans. A. E. Miller and M. C. Miller. Dordrecht-Holland: Kluwer Academic Publishers.

Polanyi, M. 1969. "Knowing and Being." In *Knowing and Being: Essays by Michael Polanyi,* ed. M. Grene, 123–37. Chicago: University of Chicago Press.

Pomian, K. 1998. "Vision and Cognition." In *Picturing Science/Producing Art,* ed. C. A. Jones and P. Galison, 211–31. New York: Routledge.

Popper, K. 1972. *Objective Knowledge.* London: Oxford University Press.

Popper, K., and J. Eccles. 1977. *The Self and Its Brain.* London: Springer International.

Price, Derek J. 1957. "The Manufacture of Scientific Instruments from c1500 to c1700." In *A History of Technology: Volume III, From the Renaissance to the Industrial Revolution c1500–c1750,* ed. C. Singer, E. J. Holmyard, A. R. Hall, and T. I. Williams, 620–47. New York: Oxford University Press.

Queraltó, R. 1999. "Technology as New Condition of the Possibility of Scientific Knowledge." *Techné* 4, no. 2.

Radder, H. 1996. *In and About the World.* Albany: State University of New York Press.

Raman, C. V. 1965. "The Molecular Scattering of Light." In *Nobel Lectures, Including Presentation Speeches and Laureates' Biographies: Physics 1922–1941,* 267–75. Amsterdam: Elsevier.

Rohrlich, F. 1991. "Computer Simulation in the Physical Sciences." In *PSA 1990,* vol. 2, ed. A. Fine, 507–18. East Lansing, Mich.: Philosophy of Science Association.

Root-Bernstein, R. S. 1991. *Discovery: Inventing and Solving Problems at the Frontiers of Scientific Knowledge.* Cambridge, Mass.: Harvard University Press.

Rothbart, D. 1997. *Explaining the Growth of Scientific Knowledge: Metaphors, Models, and Meanings.* Lewiston, N.Y.: Edwin Mellen Press.

———. 1999. "On the Relationship between Instrument and Specimen in Chemical Research." *Foundations of Chemistry* 1, no. 3:257–70.

———. 2000. "Substance and Function in Chemical Research." In *Of Minds and Molecules: New Philosophical Perspectives on Chemistry,* ed. N. Bhushan and S. Rosenfeld, 75–89. Oxford: Oxford University Press.

Rothbart, D., and I. Scherer. 1997. "Kant's *Critique of Judgment* and the Scientific Investigation of Matter." *Hyle: An International Journal for the Philosophy of Chemistry* 3:65–80.

Rothbart, D., and J. Schreifels. 2005. "Visualizing Instrumental Techniques of Surface Chemistry." In *Philosophy of Chemistry.* Boston Studies in the Philosophy of Science, ed. D. Baird, L. McIntyre, and E. Scerri, chap. 16. Dordrecht: Kluwer Academic Publishers.

Rothbart, D., and S. Slayden. 1994. "The Epistemology of a Spectrometer." *Philosophy of Science* 61:25–38.

Sabra, A. I. 1967. *Theories of Light from Descartes to Newton.* London: Oldbourne.

Salmon, W. 1994. "Causality without Counterfactuals." *Philosophy of Science* 61:297–312.

———. 1998. *Causality and Explanation.* New York: Oxford University Press.

Sarai, A., J. Mazur, R. Nussinov, and R. Jernigan. 1989. "Sequence Dependence of DNA Conformational Flexibility." *Biochemistry* 28:7842–49.

Schaffer, S. 1989. "Glass Works: Newton's Prisms and the Uses of Experiment." In *The Uses of Experiment: Studies in the Natural Sciences,* ed. D. Gooding, T. Pinch, and S. Schaffer, 67–104. Cambridge: Cambridge University Press.

Scherer, I. 1995. *The Crisis of Judgment in Kant's Three Critiques: In Search of a Science of Aesthetics.* New York: Peter Lang.

Schubert, T., F. Friedman, and H. Regenbrecht. 2001. "The Experience of Presence: Fact or Analytic Insights." *Presence: Teleoperators and Virtual Environments* 10:266–81.

Schummer, J. 1998. "The Chemical Core of Chemistry I: A Conceptual Approach." *Hyle: An International Journal for the Philosophy of Chemistry* 4:129–62.

———. 2003. "Aesthetics of Chemical Products: Materials, Molecules, and Molecular Models." *Hyle: An International Journal for the Philosophy of Chemistry* 4:77–108.

Schweber, W. 1991. *Electronic Communications Systems.* Englewood Cliffs, N.J.: Prentice-Hall.

Shapere, D. 1991. "The Universe of Modern Science and Its Philosophical Exploration." In *Philosophy and the Origin and Evolution of the Universe,* ed. E. Agazzi and A. Cordero, 87–102. Dordrecht: Kluwer Academic Publishers.

———. 1999. "Building on What We Have Learned: The Relations between Science and Technology." *Techné* 4, no. 2.

Shapin, S., and S. Schaffer. 1985. *Leviathan and the Air-Pump: Hobbes, Boyle, and the Experimental Life.* Princeton, N.J.: Princeton University Press.

Skoog, D. A., and J. J. Leary. 1992. *Principles of Instrumental Analysis.* Fort Worth: Harcourt Brace Jovanovich College Publisher.

Smith, C., and M. Norton Wise. 1989. *Energy and Empire: A Biographical Study of Lord Kelvin.* Cambridge: Cambridge University Press.

Smith, J. M. 2000. "The Concept of Information in Biology." *Philosophy of Science* 67:177–94.

Spector, M. 1965. "Models and Theories." *British Journal for the Philosophy of Science* 16:121–42.

Srinivasan, A., R. Torres, W. Clark, and W. Olson. 1987. "Base Sequence Effects in Double Helical DNA. I. Potential Energy Estimates of Local Base Morphology." *Journal of Biomolecular Structure and Dynamics* 5:459–93.

Stoney, G. J. 1871. "On the Cause of the Interrupted Spectra of Gases." *Philosophical Magazine* 41, no. 273:291–96.

Strobel, H., and W. Heineman. 1989. *Chemical Instrumentation: A Systematic Approach.* 3rd ed. New York: John Wiley and Sons.

Suckling, C. J., K. Suckling, and C. W. Suckling. 1978. *Chemistry through Models: Concepts and Applications of Modelling Chemical Science, Technology and Industry.* Cambridge: Cambridge University Press.

Swanson, J. 1966. "On Models." *British Journal for the Philosophy of Science* 17:297–311.

Taminiaux, J. 1993. "The Thinker and the Painter." In *The Merleau-Ponty Aesthetics Reader: Philosophy and Painting,* ed. G. A. Johnson and M. B. Smith, 278–92. Evanston, Ill.: Northwestern University Press.

Ter Hark, M. 1993. "Problems and Psychologism: Popper as the Heir to Otto Selz." *Studies in the History and Philosophy of Science* 24:585–609.

Toulmin, S. 1993. "From Clocks to Chaos: Humanizing the Mechanistic World-View." In *The Machine as Metaphor and Tool,* ed. H. Haken, A. Karlqvist, and U. Svedin, 139–54. Berlin: Springer-Verlag.

Ungar, E. 1996. "Mechanical Vibrations." In *Mechanical Design Handbook,* ed. H. Rothbart, section 5. New York: McGraw-Hill.

Van Brakel, J. 2000. *Philosophy of Chemistry.* Leuven, Belgium: Leuven University Press.

Vasconi, P. 1996. "Kant and Lavoisier's Chemistry." In *Philosophers in the Laboratory,* ed V. Mosini, 155–62. Rome: Editrice Universitaria di Roma.

Wallace, K. 1977. *Engineering Design: A Systematic Approach.* Berlin: Springer-Verlag.

Waller, R., ed. 1705. *The Posthumous Works of Robert Hooke.* London: Frank Cass and Co.

Warner, D. 1990. "What Is a Scientific Instrument, When Did It Become One, and Why?" *British Journal for the History of Science* 23:83–93.

———. 1994. "Terrestrial Magnetism: For the Glory of God and the Benefit of Mankind." In *Instruments,* ed. A. Van Helden and T. Hankins, 65–84. Chicago: University of Chicago Press.

Willats, J. 1997. *Art and Representation: New Principles in the Analysis of Pictures.* Princeton, N.J.: Princeton University Press.

Wilson, C. 1995. *The Invisible World.* Princeton, N.J.: Princeton University Press.

Wilson, C. T. R. 1965. "On the Cloud Method of Making Visible Ions and the Tracks of Ionising Particles." In *Nobel Lectures, Including Presentation Speeches and Laureates' Biographies: Physics 1963–1970,* 194–214. Amsterdam: Elsevier.

Winsberg, E. 1999. "Sanctioning Models: The Epistemology of Simulation." *Science in Context* 12, no. 2:275–92.

Wittgenstein, L. 1958. *Philosophical Investigations.* New York: Macmillan.

———. 1974. *Philosophical Grammar.* Ed. R. Rhees. Berkeley: University of California Press.

Woodward, J. 1989. "Data and Phenomena." *Synthese* 79:393–472.

Wright, R. 1990. "Computer Graphics as Allegorical Knowledge." *Leonardo* 23, no. 1:64–73.

Yu-tung Liu. 1995. "Some Phenomena of Seeing Shapes in Design." *Design Studies* 16, no. 3:367–85.

Zakrzewska, K. 1992. "Static and Dynamic Conformational Properties of AT Sequence in B-DNA." *Journal of Biomolecular Structure and Dynamics* 9:681–93.

Zielonacka-Lis, E. 1998. "Some Remarks on the Specificity of Scientific Explanation in Chemistry." Presentation at the Second International Society for the Philosophy of Chemistry, August 3–7, Sidney Sussex College, United Kingdom.

CREDITS

Figure 1 from M. Parsons, "Atomic Absorption and Flame Emission Spectrometry," in *Analytical Instrumentation Handbook,* ed. G. W. Ewing, 2nd ed.,p. 262. © 1997. New York: Marcel Dekker. Reprinted with permission of Marcel Dekker.

Figures 2, 3, and 11 from J. Coates, "Vibrational Spectroscopy: Instrumentation for Infrared and Raman Spectroscopy," in *Analytical Instrumentation Handbook,* ed. G. Ewing, 2nd ed. (fig. 2, p. 442; fig. 3, p. 445; fig. 11, p. 447). © 1997. New York: Marcel Dekker. Reprinted with permission of Marcel Dekker.

Figure 4 from James H. Earle, *Engineering Design Graphics: AutoCAD.* Release 12, 8th ed. © 1994. Reprinted by permission of Pearson Education, Inc., Upper Saddle River, N.J.

Table 1 from E. Ungar "Mechanical Vibrations," in *Mechanical Design Handbook,* ed. H. Rothbart, p. 35. Reproduced with permission of the McGraw-Hill Companies.

Figures 5 and 6 from *The Ecological Approach to Visual Perception* (fig. 5, p. 81; fig. 6, p. 288), by J. J. Gibson, 1986, Hillsdale, N.J.: Lawrence Erlbaum Associates, Inc. Copyright 1986 by J. J. Gibson. Reprinted with permission.

Figure 7 reprinted from *Design Studies,* vol. 16, Yu-tung Liu, "Some Phenomena of Seeing Shapes in Design," pp. 367–85, 1995, with permission from Elsevier.

Figure 8 from Constantino Baroni, *Leonardo da Vinci.* © 1956. New York: Reynal.

Figure 9 from Rudolf Arnheim, *Art and Visual Perception: A Psychology of the Creative Eye,* p. 200. © 1954 The Regents of the University of California Press.

Figure 10 from Constantino Baroni, *Leonardo da Vinci.* © 1956. New York: Reynal.

Figure 12 from 2001 edition of *Discourse in Methods, Optics, Geometry, and Meteorology* by Descartes, trans. P. J. Olscamp. New York: Hackett.

Figures 13, 14, and 15 reprinted, by permission, from Hooke, *Micrographia or Some Physiological Descriptions of Minute Bodies Made by Magnifying Glasses with Observations and Inquiries Thereupon,* 1961, Dover Publications, Inc.

Figure 16 from "Newton to Oldenburg: 7 December 1675," in *The Correspondence of Isaac Newton, Volume I: 1665–1675,* ed. H. Turnbull, p. 376. Cambridge: Cambridge University Press, 1959. Reprinted with permission of Cambridge University Press.

Figures 17 and 18 from *Principles of Instrumental Analysis, Fourth Edition,* by SKOOG. © 1992. Reprinted with permission of Brooks/Cole, a division of Thomson Learning: www.thomsonrights.com. Fax 800–730–2215.

Figures 19, 20, and 21 reprinted with permission from E. W. Müller and T. T. Tsong, "Field Ion Microscopy," *Field Evaporation,* 1969, American Elsevier Publishing Co., Inc.

Figures 22, 23, 24, and 26 from the United States Patent Office. G. Binnig and H. Röhrer, 1982. United States Patent: Scanning Tunneling Microscope. August 10, 1982. Assignee: International Business Machines Corporation, Armonk, N.Y. Patent Number: 4,343,993.

Figure 25 by permission of John Schreifels.

Figure 27 from R. Dickerson, H. Drew, and B. Conner. 1981. "Single-Crystal X-Ray Structure Analyses of A, B, and Z Helices or One Good Turn Deserved Another," in *Biomolecular Stereodynamics,* vol. 1, ed. R. Sarma, p. 6. © 1981. Published with permission from Adenine Press. http://www.jbsdonline.com.

Figures 28 and 29 from R. Maroun and W. Olson, "DNA. II. Configurational Statistics of Rodlike Chains," *Biopolymer* 27: 561–584. © 1988. Reprinted with permission of John Wiley & Sons, Inc.

INDEX

DANIEL ROTHBART is a professor of philosophy at George Mason University. As a leading scholar in philosophy of science, his areas of expertise include the philosophical aspects of research, the centrality of modeling to scientific inquiry, and the relationship between science and technology. His work has appeared in major interdisciplinary journals and scholarly volumes. He wrote *Explaining the Growth of Scientific Knowledge* and edited *Science, Reason, and Reality* as well as *Modeling: Gateway to the Unknown.*

The University of Illinois Press
is a founding member of the
Association of American University Presses.

Composed in 10.5/13 Minion
with Meta small capitals display
by Celia Shapland
for the University of Illinois Press
Designed by Dennis Roberts
Manufactured by Thomson-Shore, Inc.

University of Illinois Press
1325 South Oak Street ·
Champaign, IL 61820-6903
www.press.uillinois.edu